Exploring Physiology

For Churchill Livingstone

Editorial Director: Mary Law
Commissioning Editor: Ellen Green
Project Editor: Valerie Bain
Copy Editor: Susan J. Beasley
Production Controller: Neil Dickson
Sales Promotion Executive: Hilary Brown
Design: Design Resources Unit

Exploring Physiology

An interactive workbook for nurses

Christopher J Goodall RGN DipN RNT

Nurse Teacher, North Yorkshire College of Health Studies, York, UK

Illustrations by Robert Britton

CHURCHILL LIVINGSTONE
EDINBURGH LONDON MADRID MELBOURNE NEW YORK AND TOKYO 1994

CHURCHILL LIVINGSTONE
Medical Division of Longman Group UK Limited

Distributed in the United States of America by
Churchill Livingstone Inc., 650 Avenue of the Americas,
New York, N.Y. 10011, and by associated companies,
branches and representatives throughout the world.

First published 1994

ISBN 0-443-048177

British Library Cataloguing in Publication Data
A catalogue record for this book is available from the British
Library.

Library of Congress Cataloging in Publication Data
A catalog record for this book is available from the Library of
Congress.

The
publisher's ·
policy is to use
**paper manufactured
from sustainable forests**

Produced by Longman Singapore Publishers Pte Ltd
Printed in Singapore

Contents

Preface

Physiology is a vital component of many health care courses, both pre- and post-registration, yet it is a subject with which some students may be unfamiliar at the end of their school careers. Again, health care courses attract mature students, women and men who may not have done an academic course for many years.

It is common, therefore, for teachers to face students in the early stages of their course with widely differing levels of knowledge in human physiology. This Workbook will help students and teachers to work together, at a pace suited to individual students, in order to reach a common level of understanding. The text aims to be informal and encouraging, as well as informative. It is based closely, though not exclusively, on Churchill Livingstone's new textbook *Physiology and Anatomy: A Basis for Nursing and Health Care* by Sigrid Rutishauser, to whom the author would like to express his sincere thanks for her encouragement and hard work in looking through the manuscript. Students will find the combination of the Workbook and Rutishauser a formidable yet readable learning resource.

This book is for all the students, past and present, and the staff of the North Yorkshire College of Health Studies. It is also dedicated to the memory of Phyllis Goodall who was, for many years, both nurse and teacher.

1994   C.J.G.

Introduction for students and teachers

The principal purpose of this Workbook is to provide an interactive text which is compact and relatively inexpensive. Its aim is to help student nurses gain a basic understanding of physiology; consequently it will be of most use to students in the first 9 to 12 months of their Common Foundation Programme. However, it should also prove valuable for certain of the Branch Programmes, and for nurses coming back into the profession after a break of some years.

There are two important questions relating to physiology in the context of nursing:

1. Why is the study of physiology important for the student nurse?
2. How might physiology best be taught?

WHY STUDY PHYSIOLOGY?

Physiology, though a fascinating subject in its own right, informs nursing practice. The discussion document for Project 2000 (UKCC 1986) talked of 'the knowledgable doer' and with this sentiment I agree. Physiology is not a subject that a nurse learns exclusively in the classroom. While basic facts and concepts are found in textbooks and lecture theatres, the living subject is there in front of the nurse, each time he or she cares for a patient.

Safe nursing care is based on a good grounding in physiology. For example, if a student has learned the main functions of the red blood cell, it becomes possible for that student to appreciate the physiological effects of a reduced red cell count or haemoglobin level. The student, later observing a patient who is very pale and who finds the least exercise exhausting, can see in real life these physiological effects. For that student, the nursing care devised for the anaemic patient may be seen as well grounded in physiology. He or she has become a 'knowledgable doer'.

LEARNING PHYSIOLOGY

Learning that is an active process is likely to be more efficient, and more enjoyable, than learning that is mostly passive. For this reason, I have rejected a simple question and answer format because, however cunningly answers to lists of questions are hidden within the covers of a book, the temptation is there for a student to skip from one to the other. This entails little 'learning action' beyond flipping through pages. Instead, readers of this book must seek out information for themselves; they must be active rather than passive in their learning.

Rather than providing answers, there are frequent, often precise, references to physiology texts (but please see the note at the end of this Introduction). Consequently the student must seek out relevant information from those texts. Very soon, each student will feel able to decide which of these books is most appropriate for him or her and the course being undertaken; which text the student is most comfortable with. At the end of this Introduction I give a list of physiology texts referred to in this Workbook, with a brief comment on some of them.

As regards references, I would always urge the reader to make use of a number of books. This is particularly important for complex topics, where reading the same subject in a second or third text may provide additional enlightenment.

AIM OF THIS BOOK

This Workbook is designed to be used with other physiology texts. It does not usurp the position of the nurse teacher as overall facilitator of students' learning, nor is it my intention that this book should be used as a replacement for formal teaching. Nor would I suggest that this book be used unswervingly as the physiology basis of a curriculum. It should be utilised, in part or as a whole, within a planned learning programme. It is for the teacher and student to negotiate how much use might be made of this book. I would be happy if teachers constructed their own guided studies for students, using either the format or certain ideas provided by this book.

It will be especially valuable where there are large classes consisting of students with widely differing backgrounds in anatomy and physiology – frequently a problem with nursing intakes – for it will encourage individuals to progress at their own rates, with the aim of achieving a common level of understanding.

STRUCTURE OF THIS BOOK

This book is set out using the familiar 'systems of the body' approach, with the systems grouped together where appropriate into Units.

Throughout the Workbook there are four different 'Learning elements'. The three commonest elements are:

- Work Sheets
- Guided Studies
- Check Quizzes.

Work Sheets are relatively short, often providing an overall view of the particular body system being studied. They can be used as a method of achieving a level of background knowledge across a student group. Students who have studied biology at school will probably need simply to skim through these Work Sheets, using them as a springboard for further study. Students who entered nursing with no formal qualifications in biology will need to spend much more time on these preparatory Work Sheets.

It is difficult to provide a precise time limit for the completion of these Work Sheets, but about 3 hours may be appropriate – a morning or afternoon. Some students may need a full day as well as close tutorial support.

Many body systems are covered, following an initial Work Sheet, with a much more detailed **Guided Study**. These begin where the preliminary Work Sheets leave off, so you will appreciate how important it is for the non-biologically qualified student to attempt the easier Work Sheet first.

Again, timing is difficult to predict, but perhaps 2 'college days' (with, say, 6 hours of work per day) will probably be sufficient. Reference books and tutor support will again be vital. Both Work Sheets and Guided Studies contain suggestions for places to take a break.

You will find that the literary style of these Guided Studies and Work Sheets is informal without, I hope, being chatty. I tend to use them in my own teaching for the students I know, and I find it easy to 'talk to' my students through the printed page. I don't believe that written learning methods should necessarily be more formal than oral teaching sessions.

Each Unit includes at least one short **Check Quiz**, the third learning element, devised to help students assess the knowledge they have gained from following the Unit's previous learning elements. These Check Quizzes are sometimes built around brief patient profiles and nursing situations (for example, a lady with severe anaemia who requires help with washing and dressing). In this way I hope to underline the importance of physiology to sensitive and appropriate nursing practice.

The fourth learning element is the occasional **Suggested Reading**. This text cannot hope to include all possible areas of physiology necessary for the student nurse following a Common Foundation Programme, though I have attempted to cover the most important. Rather than omit a particular topic completely because of lack of space, however, I have put together some fairly detailed references which aim to provide reasonably full coverage. Inevitably, though, these lists of references are less interactive than the other learning elements in the Workbook.

ICONS

To help you find your way around the Workbook there are repeated icons denoting, for example, appropriate places to take a break, or a commentary on certain questions. These commentaries give additional 'clues' to help you tackle complex questions. They take the place of the simple – and non-interactive – provision of answers, as I've already discussed.

Sometimes student activities are suggested, highlighted by their own icon. These entail simple exercises like taking a colleague's pulse or blood pressure, and they're used to emphasise a particular physiological point.

You will find that subjects for further study readily suggest themselves as you progress through this book – and through your nursing course, such is the fascination of physiology. Thinking, 'Yes, but I wonder why . . .' is a healthy response to the textbooks you read. So, interspersed throughout both Work Sheets and Guided Studies there are suggestions for further study, shown by their own icon.

APPROPRIATE DEPTH

A problem that seems commonly to arise with any learning material is: 'To what depth do I need to study this?' Students frequently ask this question, not from idleness (hoping to get away with the minimum of study) but with a genuine concern for the level of understanding they need to attain.

A good answer lies in that all-important word, 'understanding', and it explains why I have often provided Work Sheets first, giving a fairly basic overview of a body system before going into greater detail with a subsequent Guided Study.

Students should go into as much detail for a particular subject as they understand. Two or three simple sentences that provide a comprehensible idea of the subject being studied, helped by a clear diagram, are more valuable than many paragraphs of impressive looking, detailed information copied wholesale from a textbook. As a student you will need to make frequent notes from textbooks, but copying is much less useful than paraphrasing: sifting the information in front of you until you arrive at the 'meat' of the text, which you can then express in your own words.

As well as seeking the guidance of your teachers in fixing the appropriate depth for which to aim, remember that your notes are for *your* benefit, and are designed with *yourself* in mind. Ultimately, they will assist you in your nursing care, and your understanding and organisation of that care.

This book should be regarded as one learning resource among many – the others including your preferred physiology texts, articles in journals, your teachers and colleagues, and the patients you nurse.

REFERENCE
UKCC 1986 Project 2000: a new preparation for practice. UKCC, London

PHYSIOLOGY TEXTS REFERRED TO IN THIS WORKBOOK
The two textbooks referred to most frequently in this Workbook are:

Rutishauser S 1994 Physiology and anatomy: a basis for nursing and health care. Churchill
Livingstone, Edinburgh
(A new, detailed physiology text, well illustrated, and with many nursing applications
which help relate physiology to client care and nursing intervention.)
Wilson K 1990 Ross & Wilson, Anatomy and physiology, 7th edn. Churchill Livingstone,
Edinburgh
(Perhaps the classic textbook for nurses, it gives more emphasis to anatomy than to
physiology. It is very well illustrated, and remains an excellent textbook.)

To obtain the most from this Workbook, you will certainly benefit from having a copy of
Rutishauser with you.
Other physiology texts:

Bursztyn P 1990 Physiology for sportspeople. A serious user's guide to the body. Manchester
University Press, Manchester
(By studying how the body behaves during exercise, we can gain a clearer understanding
of 'normal' physiological processes. This book provides useful information, and some
helpful cartoon-like diagrams.)
Hinchliff S, Montague S 1988 Physiology for nursing practice. Baillière Tindall, London
(A very detailed physiology text, which demonstrates the close relationship between
physiology and nursing situations.)
Hubbard J, Mechan D 1987 Physiology for health care students. Churchill Livingstone,
Edinburgh
Jennett S 1989 Human physiology. Churchill Livingstone, Edinburgh
(Both of these texts are detailed, and somewhat technical in their approach, but are useful
as additional reading.)
Marieb N 1992 Human anatomy and physiology, 2nd edn. Benjamin/Cummings, Redwood
City, California
(Contains excellent coloured diagrams, and fascinating electron micrographs.)
Mackenna B, Callander R 1990 Illustrated physiology, 5th edn. Churchill Livingstone,
Edinburgh
(A book that explains complex physiological processes by means of detailed diagrams. It is
most useful perhaps as a supplementary text to others.)

Two other physiology texts, by the same author, I include here because they have long been
my favourites. Despite their detail I find the writing is clear and precise, and without the
forced informality which some American texts adopt. Beware, however, of some of the
Americanised spelling, such as edema and hemoglobin.

Guyton A 1984 Physiology of the human body, 6th edn. Holt Saunders, Philadelphia
(Despite its comparative age, this edition is still available in British book shops.)
Guyton A 1991 Textbook of medical physiology, 8th edn. W B Saunders, Philadelphia
(A weighty and expensive volume; but don't be frightened to dip into its pages in order to
gain additional understanding, or to pursue a topic further which especially interests you.
Most nursing and medical libraries should have a copy of this edition or its predecessor.
Page references given in this Workbook, however, refer to the 8th edition.)

The occasional additional references, for example to texts on pharmacology or medical
nursing, are provided in full at the end of the Work Sheet or Guided Study in which they
occur.
Note: Most physiology texts are updated at regular intervals, and consequently some of the
references in this Workbook to precise pages or figures in those texts may soon be out of
date. However, you will be able to find the editions of texts mentioned here in your nursing
college libraries; or you can find the appropriate passage in an updated edition. Please note,
therefore, the year of publication given for each of the texts referred to in this Workbook, for
this will guide you to the appropriate edition.

Support and movement

Contents

UNIT 1

■ The cell (work sheet)

Time for completion   about 6 hours (or 1 'college day')

Overall aim   To review the structure of the 'general cell', the work performed by each of its parts, and the role it plays in the community of cells that is the human body.

Introduction   It is logical to begin our study of physiology by reviewing the structure and functions of cells, because the human body is made up of many millions of these building blocks. Cells vary greatly in shape and function, just like building materials. If we were constructing a building together, we'd realise that our design of building would greatly influence the type of materials we could use. So the shape and function of cells making up the different parts (tissues) of the body dictate the job performed by those tissues.

For example, a cell that secretes a chemical (such as a hormone) will look quite different from a nerve cell that conveys an electrical message from toe to spinal cord. How long do you think such a nerve cell might be in an adult?

In this Work Sheet we examine the structure and function of a 'general cell'; in other words we'll concentrate on those structures and functions that tend to be the same for most types of cells.

Background questions   1. Find a diagram showing cells that form the lower layers of the skin. (See, for example: *Wilson 1990, Fig. 11.1; Rutishauser 1994, Fig. 15.2*). Draw just one of these cells. Now add a diagram showing one example of a type of white blood cell called a lymphocyte (*Wilson 1990, Fig. 4.8; Rutishauser 1994, Fig. 4.12*). Label the following cellular features in each of your drawings:

- cell membrane (or plasma membrane)
- cytosol
- nucleus
- nuclear membrane.

This will help to highlight the similarities of these cells you've drawn. If you need help, look at *Wilson 1990, Fig. 2.1* or *Rutishauser 1994, Fig. 2.3*.

2. Now find diagrams of a red blood cell (erythrocyte) and a nerve cell (*Wilson 1990, Figs 1.3 and 12.2; Rutishauser 1994, Fig. 2.9*). Add these to your notes pointing out their obvious differences from the cells you drew earlier, and also their similarities.

Specific questions   **THE CELL AND ITS ORGANELLES**
Time for completion: about 1½ hours

Within the 'general cell' lie many different structures, referred to as organelles. Not all of these organelles are present in all types of cell. You've already found that erythrocytes lack a nucleus, for example.

1. Make brief notes on each of the following organelles:

- endoplasmic reticulum (rough and smooth)
- Golgi apparatus
- lysosomes
- peroxisomes
- nucleus
- nucleolus.

(Note that mitochondria are left till Question 2.)

You should write no more than one short paragraph for each, concentrating on attaining a clear overview of their functions. The following references are suggested: *Wilson 1990, pp. 15–17; Hubbard & Mechan 1987, pp. 6–10,* (includes some interesting notes on the light and electron microscopes); *Jennett 1989, pp. 25–27; Rutishauser 1994, pp. 21–22; Guyton 1984, pp. 18–21; Marieb 1992, pp. 76–85,* (particularly recommended for its excellent diagrams and electron micrographs).

As with all listed references in this Workbook, I've given those above in order of increasing complexity. You do not have to work through every one, but I'd suggest that it's useful to study each topic using at least two textbooks.

 COMMENTARY ON QUESTION 1

Sometimes I find that diagrams in textbooks contain far too much information to take in at once. Copying such a diagram into your notes is of limited value and I'd suggest that you build up your understanding by means of 'series diagrams', which are a feature of this Workbook.

Start by drawing a large version of Figure 1.1, consisting of the barest outline of the cell's structure. Supposing that you are going to make notes on lysosomes, draw them first in a smaller version of Figure 1.1 (see Fig. 1.2). Then, when you're clear about their appearance and function, add lysosomes to your original large-scale 'master diagram' of a cell.

Do the same for the next organelle you study. Make a small diagram first containing that particular organelle, then add it to your master diagram. In this way, you build up your knowledge as you construct your diagram. In each of the small-scale 'series diagrams' label only the organelle you're concerned with.

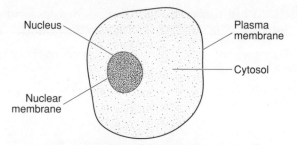

Fig. 1.1 Basic structure of the cell.

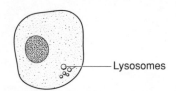

Fig. 1.2 Lysosomes – first of a 'series diagram'.

2. Make notes on another organelle, the mitochondrion. Mitochondria are sometimes called the power houses of the cell. Why is that?

Make brief notes on the formation of ATP and its significance for cellular function. What happens when ATP splits to form ADP? How is ATP formed in the first place? Don't get bogged down with the structure of these molecules, but concentrate on the broad view of their function within a cell. Later on we'll see the significance of energy creation within a cell when we look at how substances cross the cell's membrane. However, at the moment can you think of some cells that have to expend energy in order to function? See: *Marieb 1992, pp. 57–58; Rutishauser 1994, pp. 211–213; Guyton 1984, pp. 37–38; Guyton 1991, pp. 19–21.*

(As in future Work Sheets and Guided Studies, references to Guyton 1991 are provided for further reading, and not as primary study material.)

THE CELL AND ITS PLASMA MEMBRANE
Time for completion: about 1½ hours

Membranes enclose the cell, its nucleus, and many of its organelles. In this part of the Work Sheet we'll look at the cell membrane (plasma membrane) only. The plasma membrane provides some degree of protection for the cell. It defines its shape (though some cells can change shape). It allows into the cell some substances while keeping out others.

The plasma membrane consists mostly of proteins and phospholipids, with a small amount of carbohydrates. Looking at a diagram of the membrane, one obvious feature is the double layer (bilayer) of lipids. See: *Hubbard & Mechan 1987, Fig. 1.8; Rutishauser 1994, Fig. 2.2; Guyton 1984, Fig. 2.5; Marieb 1992, Fig. 3.2.*

3. Copy Figure 1.3, which shows just a few molecules of the phospholipid bilayer. Add labels showing which part of the bilayer is hydrophilic, and which is hydrophobic. What do these terms mean?

4. Now we turn to the protein elements in the plasma membrane. Make brief notes on:

   - integral proteins
   - peripheral proteins.

Draw a small diagram of each of these, then add them to your original large-scale diagram. Figure 1.4 shows how I've added an example of an integral protein to my diagram. You should draw something similar, then go on to add an example of a peripheral protein.

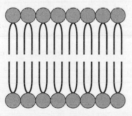

Fig. 1.3 Lipid bilayer. On your diagram, label the hydrophilic and hydrophobic parts.

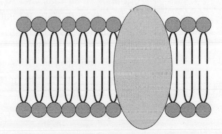

Fig. 1.4 Lipid bilayer with integral protein added.

The figures referred to above will help you with Question 4. You might find the following passages of text helpful too: *Hubbard & Mechan 1987, pp. 10–12; Rutishauser 1994, p. 16; Guyton 1984, pp. 25–26; Marieb 1992, pp. 63–66.*

Your notes should be brief at the moment, keeping simply to the structure and appearance of these proteins. We'll be dealing with how substances pass through the plasma membrane, and the role of proteins in this, next.

FURTHER STUDY

You may wish to read more about the chemical composition of lipids and proteins. Try one or more of the following: *Rutishauser 1994, pp. 9–16; Guyton 1984, pp. 23–24; Marieb 1992, pp. 45–53.*

5. We've already said that the plasma membrane keeps out some substances while letting in others. Here we look at how this is achieved.

There are two broad methods by which substances cross the membrane: those that are passive and those that are active (the latter involving the expenditure of energy by the cell). You should look at:

- simple diffusion
- facilitated diffusion
- osmosis
- active transport.

Try also to discover the meaning of the terms exocytosis and endocytosis.

 COMMENTARY ON QUESTION 5

Rutishauser 1994, p. 16, provides a brief introduction to the role of proteins in cell membranes in the passage of substances. *Hubbard & Mechan 1987, pp. 12–14* gives a little more detail. *Marieb 1992, pp. 66–74*, gives far more detail and several excellent diagrams.

If you haven't met the terms diffusion and osmosis before, see *Wilson 1990, pp. 45–46*. In simple diffusion, substances in high concentration (for example salt) will cross a semipermeable membrane into a weaker solution. The greater the difference in concentration (i.e. the steeper the concentration gradient) the sooner the two concentrations equalise. You may like to express both diffusion and osmosis in diagrams of your own.

A concentration gradient is quite easy to visualise. Remember, though, that in order for substances to cross from a higher concentration to a lower, the membrane must allow those substances to pass. This is where the proteins in the lipid bilayer of the plasma membrane come in. Some substances are simply allowed through tiny channels; some are transported across the membrane, even against a concentration gradient (i.e. going 'uphill').

6. Earlier it was said that a relatively small part of the plasma membrane consists of carbohydrates, in the form of glycoproteins. These act as cell markers, and play a part in the cell recognising other substances and being itself recognised. Make a very brief note about the function of glycoproteins in the plasma membrane, and add an example to your diagram of a cell membrane. See: *Guyton 1984, p. 26* (a very brief description); *Rutishauser 1994, Fig. 2.5D and p. 16; Marieb 1992, pp. 75–76.*

 This is a good place to take a break.

NUCLEIC ACIDS, PROTEIN SYNTHESIS AND CELL DIVISION
Time for completion: about 2 hours

This highly complex subject will be covered briefly here, but with suggestions for further study. The aim of this section is to help you gain a clear, basic idea of the role of nucleic acids in the functioning of the cell. The groundwork achieved here will, I hope, assist your future, more detailed study.

If the bridge of a ship can be regarded as its control centre, so we can call the nucleus the control centre of the cell. The nucleus controls its cell's functions – producing proteins for construction work (e.g. building the cell's organelles) and proteins for enzymes which play a part in chemical reactions.

7. First of all, find out what the abbreviations DNA and RNA stand for. For simplicity, I'll stick to the abbreviations in this Work Sheet.

8. DNA inside the nucleus of a cell forms RNA which diffuses out into the cytoplasm (via pores in the nuclear membrane, which is much more 'leaky' than the plasma membrane). In the cytoplasm, this RNA helps in the manufacture of cellular proteins. I've tried to express this in a simple diagram (Fig. 1.5). Add labels to your own version of this diagram, showing which part is the nucleus, and which the cytoplasm.

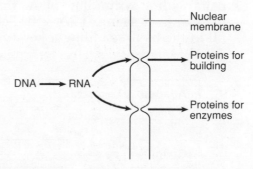

Fig. 1.5 DNA controlling cell function. On your diagram, label the nucleus and cytoplasm.

9. Now we look at the substances that make up DNA and RNA. First, take a look at a diagram of the double helix – a twisting double-strand of DNA. We're going to build up an understanding of this complex structure by making another 'series diagram'. Diagrams of the double helix are to be found in: *Wilson 1990, Fig. 2.3; Rutishauser 1994, Fig. 2.6C; Guyton 1984, Fig. 4.2; Marieb 1992, Fig. 3.24* (this shows the DNA double helix splitting apart, rather like a zip being undone).

10. Both DNA and RNA are built up of many nucleotides, which are themselves made of these components:

    • base
    • sugar
    • phosphate.

List the five different bases, which are divided into:

    • two purines (present in both DNA and RNA)
    • three pyrimidines (one unique to DNA, one unique to RNA, one present in both).

Name the sugar that is part of DNA, and the sugar that is part of RNA. You'll now see why DNA and RNA are so named.

11. The double helix looks rather like a rope ladder that's been twisted on itself. To start our series diagram, draw an untwisted version of the ladder, with its outer edges made up of alternating phosphates and sugars, as in Figure 1.6.

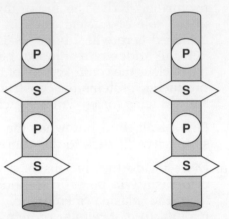

Fig. 1.6 Beginning of the DNA 'rope ladder'. P = phosphate; S = sugar.

12. The 'rungs' of the ladder are made of pairs of bases, in which a purine always pairs with a pyrimidine. Show this complementary base-pairing by filling in the blanks in Figure 1.7A & B. Then in Figure 1.8 (the complete rope ladder) fill in the blanks, adding a key to explain the symbols used.

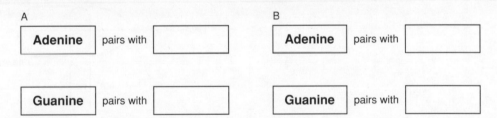

A

Adenine	pairs with	

Guanine	pairs with	

B

Adenine	pairs with	

Guanine	pairs with	

Fig. 1.7 Complementary base-pairing in (A) RNA and (B) DNA. Fill in the blanks in your diagram.

In your imagination, grab hold of the bottom edge of your 'ladder' and twist it – and you have the double helix (which is very difficult to draw). Check your imagined helix with the diagram in your textbook.

 FURTHER STUDY

In this Work Sheet I've used symbols for the bases, phosphate and sugars. Chemical formulae for these are given in Rutishauser 1994 and Guyton 1984. You'll be able to discover why the bases (the 'rungs' of the ladder) form bonds which can both hold together and split apart.

How does DNA within the nucleus affect what happens in the cytoplasm? After all, it's in the cytoplasm that much of the cell's activity occurs. In brief, DNA inside the nucleus gives rise to RNA, which makes its way through the nuclear membrane into the cytoplasm. We now look at how this occurs.

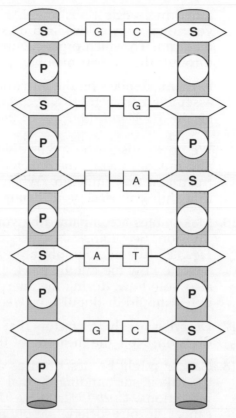

Fig. 1.8 Complementary base-pairing in DNA. On your diagram, add a key to the symbols used, and fill in the missing symbols.

13. Read what *Rutishauser 1994, p. 19*, has to say about the genetic code. Remember that the DNA in our cells gives rise to the many complex proteins that lead to our physical characteristics. Make notes on the terms gene and codon, and on how codons represent specific amino acids. Draw a diagram of a length of DNA with the codon(s) for one or two amino acids.

 COMMENTARY ON QUESTION 13

Remember that each of the bases on the DNA helix is automatically paired with another: adenine with thymine; guanine with cytosine. But the bases down one side of the DNA 'ladder' occur in many different sequences. A sequence of three bases (with their links) is a codon – that is, the code for one amino acid. Thus the code for glycine is GGG, or three occurrences of guanine. And guanine is always linked, across on the other side of the ladder, with which base?

This is complex; perhaps you've heard of Morse code, an early form of signalling in which a series of dots and dashes represents each letter of the alphabet. These dots and dashes could be flashes from a torch, or tones from a radio transmitter. Just as the letter S is represented by three dots, so an amino acid code consists of three bases in a particular sequence.

14. For these codes to get into the cell cytoplasm, the twin strands of DNA have to split apart (like a zip opening) for a short distance, and a version of RNA,

called messenger RNA (mRNA), is formed. See *Rutishauser 1994, Fig. 2.7*. Make your own copy of this, and make notes to explain what is happening in your diagram. To which organelle in the cytoplasm does mRNA pass, after going through the nuclear membrane?

15. Now make notes on the formation of transfer RNA, and how amino acids are assembled from the RNA coding. See: *Rutishauser 1994, p. 19; Guyton 1984, pp. 48–51; Marieb 1992, pp. 91 and 95–98*.

We now consider how cells divide. They do so at greatly differing rates. Skin and mucous membrane cells tend to be replaced quickly because of the rough treatment they undergo. (When you're on the wards, count how many times in a shift you wash your hands.)

16. Make notes accompanied by your own 'series diagram' on the phases of cell division. See: *Rutishauser 1994, Fig. 2.10; Guyton 1984, Fig. 4.12; Marieb 1992, Fig. 3.25* (the micrographs are especially interesting here).
    Note how the mitotic phase is much shorter than the cell's interphase, but also note how, during the interphase, DNA is preparing for replication. Your notes should distinguish between mitosis and cytokinesis.

17. Describe briefly the factors that influence (i.e. either enhance or prohibit) cell division. Note the meaning of the term contact inhibition.

18. Cancer might be described as a collection of cells that divide more rapidly than their surrounding normal cells, and which are not restricted by contact inhibition. *Marieb 1992, pp. 90–91 and 94*, provides an interesting description of cancer. In our society, we often use words other than cancer – tumour, for example, or growth. Find out the physiological differences between malignant and benign tumours. You probably already know that a malignant tumour is more likely to spread to other parts of the body, but does this mean that a benign tumour is necessarily harmless? Discuss this point with your colleagues.

■ Tissues of the body (suggested reading)

The human body is rather like a complex human society, because each has its many different parts contributing to the functioning of the whole. Think, for example, of how many human beings must have contributed to the production of this Workbook; and how many people work together, probably without knowing each other, in different parts of the world to produce the food you eat in a single meal.

Similarly, the human body is a highly complex structure. Its cells in the skin look different, and behave differently, from cells in the brain; and though they look similar to cells of the bladder wall, there are important differences.

Differentiation is the term given to the way in which cells develop differences in appearance and function. What distinguishes, cellularly, the human from the amoeba is not just that the human consists of many millions of cells and the amoeba of one cell only; it's also that human cells are greatly differentiated. See: *Mackenna & Callander 1990, p. 10; Hinchliff & Montague 1988, p. 16.*

Tissues consist of collections of many cells, each tissue's cells being similar in appearance and function. It's usual to group human tissues into four main divisions (each with its subdivisions):

- epithelial
- connective
- nervous
- muscle.

I'm going to suggest some reading that will introduce only some of the many different types of tissue. First, then, comes the list of possible subject areas:

1. Epithelial tissue

   - stratified epithelium
   - transitional epithelium
   - mucous membrane.

2. Connective tissue

   - fibrous tissue
   - areolar tissue
   - adipose tissue
   - cartilage
   - elastic tissue.

(This division also includes the blood; and it might be a surprise to discover that blood is regarded as a tissue. Yet it is, after all, many cells contained within a fluid environment. Blood is covered in Unit 3 of this Workbook.)

3. Muscle tissue

   - skeletal muscle
   - smooth muscle
   - cardiac muscle.

(Here I'd suggest you try to gain a clear understanding of the similarities and differences in these three types of muscle. It's obvious where cardiac muscle and skeletal muscle are found: but where is smooth muscle found?)

4. Nervous tissue

   - nerve cell.

(Unit 2 goes into more detail about nervous tissue.)

Now here are some passages of suggested reading, as usual in ascending order of complexity:

Mackenna & Callander 1990, pp. 11–19 (a useful summary, but I wouldn't suggest it as your main text)
Wilson 1990, pp. 18–26
Hubbard & Mechan 1987, pp. 23–28 (epithelium); *pp. 29–40* (connective tissue); *pp. 40–41* (muscle); *pp. 51–53* (nervous tissue) (parts of Ch. 1 are too complex for this stage of the Workbook, and have been omitted)
Marieb 1992, Ch. 4 (provides excellent diagrams and some very useful electron micrographs).

■ The skin (work sheet)

Time for completion	About 4 hours
Overall aim	To relate the structure of the skin to some of its functions: protection of the body; wound healing; and temperature regulation.

Introduction

The skin is perhaps the most visible organ of the body, depending on the amount of clothing worn. Together with the muscles and skeleton, it forms an important part of one person's physical attractiveness to another. The receptive nature of the skin – our ability to feel texture and pressure via its nerve endings – is important in making us aware of our environment and assessing its danger.

Sexual activity of the human relies greatly on the sensitivity of the skin. It provides sensations and receives touch; it reveals inner feelings (by blushing, for example). In some adolescents, skin blemishes can cause tremendous misery and social isolation.

In some parts of the world, skin colour helps or hinders a person's social, economic, and academic progress throughout his or her whole life. In such 'civilisations' a person is judged primarily by skin colour. Fingerprints – those unique patterns on the tips of our fingers – play an important part in the fight against crime by, in some instances, identifying a person's presence at the scene of the crime.

The skin plays a part in the fluid balance of the body, by secreting fluid in the form of perspiration. We've probably all experienced how the amount we perspire varies according to temperature and humidity, and the amount of physical exercise we do. Much of the time we may be completely unaware of fluid loss from our skin.

In this Work Sheet we look at some of the many functions of the skin. Because of this diversity there are cross-references to other Units in this Workbook.

Background questions

These initial questions help you gain a basic understanding of the skin's structure, on which you can build in the following specific questions.

1. In your preferred physiology textbook find a diagram showing a cross-section of the skin, and distinguish between the epidermis and the dermis. In which of these layers are blood vessels and nerve endings mostly found?

2. Thinking back to the preceding Suggested Reading on tissues of the body, of what type of tissue is the skin composed?

3. How does the shape of cells in the epidermis change as they near the surface of the skin? Show these changes in a simple diagram. Label the different layers.

4. Draw a diagram of the skin by using the 'series diagram' technique I've suggested before in this Workbook. First draw your main diagram showing the epidermis (with its layers) and the dermis. Then add:

   - nerve endings
   - blood vessels
   - sweat glands
   - hairs, hair roots and sebaceous glands.

   The idea is for you to end up with a diagram like *Fig. 11.2* in *Wilson 1990*. But to prevent your diagram becoming too complex, draw alongside it a further small-scale diagram of each component that you add.

Specific questions

THE SKIN AS PROTECTIVE LAYER
Time for completion: about 1 hour

1. What happens to the structure and composition of the cells as they move from the lowest layer of the epidermis towards the surface? In your description of these changes, ensure that you explain the term keratinisation.

2. How are finger- and toenails formed, and what is their function?

3. Make notes on melanin, its presence in the skin, and its protective role against ultraviolet light.

 FURTHER STUDY

Have you read of any recent changes in the atmosphere that have led to an increased threat to humans from sunlight? Whereabouts on the globe is this threat greatest, and what skin colour is most at risk?

---

4. Note the presence in the dermis of nerve endings that are stimulated by:

   - light touch
   - pressure
   - heat.

   Write a brief paragraph about the protective function of these nerve endings, bearing in mind that we'll be studying the nervous system in more detail in Unit 2.

 COMMENTARY ON QUESTION 4

Stimulation of certain nerve endings can lead to muscular movement. This can be an involuntary, or reflex, response, such as dropping a very hot plate; or a voluntary response as, for example, when a class of students begins to shuffle towards the end of a long lecture. It may not be that the lecture is boring; more probably the tissues pressing on hard seats cause discomfort and prompt the students to move.

People who have fractured spines lose sensation below the level of the injury. They are unable to feel pain, pressure, heat or cold in their legs and buttocks (and sometimes higher). Knowing that to sit for a long period in one position can cause tissue damage, such as a pressure sore, how might a young woman in her wheelchair prevent such damage from occurring? What action might she take herself?

---

5. As epidermal cells approach the surface of the skin, they become flattened, and eventually flake off. What is this process called? The shedding of skin cells occurs particularly during washing, dressing and undressing. The skin surface is usually occupied by colonies of microorganisms, and these too are shed. They collect in a patient's bed, on the sheets and pillow cases. Discuss with your fellow students how, when making or changing beds on a ward, you might limit the numbers of microorganisms introduced into the atmosphere. You could also think of films you've seen of operating theatres (assuming

you haven't yet been to theatres in your nurse training). What steps do theatre staff take to prevent their own skin cells entering a patient's open wound?

6. Describe how and where the substance sebum is produced, and its role in helping the skin protect the body by keeping out foreign microorganisms.

 COMMENTARY ON QUESTION 6

The unbroken skin provides an efficient barrier to most germs and other foreign agents, such as chemicals. Bacteria and viruses usually gain entrance to the body (i.e. to the blood) through a cut or a graze. A surgical wound, or a scraped knee, provide excellent routes for germs to enter the blood (hence all the efforts in a theatre to keep the environment germ-free). Think also about conditions that might cause one's skin to split; is this likelier to happen to dry skin or to soft, moist skin? Now, what role does sebum play in preventing cracking of skin?

7. How 'waterproof' is the skin? Describe its normal reaction to water, and what happens when it is soaked for a long period. (We've all noticed the appearance of our skin if we've stayed in the bath too long – but what has happened at a cellular level?)

To help with the above questions, refer to: *Wilson 1990, Ch. 11; Rutishauser 1994, Ch. 15; Marieb 1992, Ch. 5.*

 Have a short break here.

WOUND HEALING
Time for completion: about 1 hour

In the previous section I mentioned surgical wounds. The very nature of such a wound is that it is a neat cut, made with a very sharp scalpel (and one that is sterile) and, usually, with its edges held together with sutures (stitches) or clips. This type of wound will heal more quickly than one where there is a lot of tissue loss (e.g. a deep pressure sore), and where the edges are jagged and are too far apart to come together, as in some severe accidents (e.g. explosions or road accidents).

The first type of wound, the neat surgical cut, is said to heal by primary intention (primary healing). The larger, more jagged cut, with more tissue loss, heals by secondary intention (secondary healing).

8. Draw your own 'series diagrams' illustrating primary healing. Note the position of the blood clot, and how it eventually disappears. See how the top surface of the skin, the epidermis, eventually bridges the gap formed by the cut.

Note the role of certain blood cells called phagocytes. If this term is new to you, look it up in a nursing or medical dictionary, or turn to *Wilson 1990, p. 52.* Phagocytes are one of the white blood cells discussed in Unit 3.

9. Now draw another 'series diagram' showing secondary healing. Make sure your diagrams show a much wider wound with a bigger blood clot. Note the

role played by granulation tissue. How long does secondary healing take, approximately, compared with primary healing?

Suggested reading: *Wilson 1990, pp. 237–238; Hinchliff & Montague 1988, pp. 575–576.*

10. The second of these references also briefly describes a number of factors that can delay healing. One of these is wound infection. You'll find a description of the skin's normal flora (i.e. microorganisms that exist naturally on the skin) in *Hinchliff & Montague 1988, pp. 572–573.* If the skin isn't cleaned properly before a surgical incision is made, what might be the result?

 FURTHER STUDY

At probably a later stage in your nurse training, you'll learn how to remove clips or sutures from a surgical wound that's properly healed. If this is done thoughtlessly, microorganisms can be introduced below the surface of the skin. For example, a length of suture material can be drawn through the skin as it is pulled out. You'll be shown how to avoid this, by cutting the suture in a certain place. In your physiology notes, you may like to leave a space so as to add this later information on avoiding wound infection. This would perhaps be a good place for these additional notes, near your information on skin normal flora.

 Have a short break before the final part of this Work Sheet.

THE ROLE OF THE SKIN IN TEMPERATURE REGULATION
Time for completion: about 1½ hours

Not all the changes made by humans in response to severe temperature swings in weather are physiological. Unlike birds and mammals, humans cannot fluff up their feathers or fur in response to the cold. Nor do we hibernate. We do, however, construct the human equivalent of a burrow in which to shelter from the cold, by building houses with insulated walls and windows and central heating. And, like adders sunning themselves on a bare patch of earth, we also enjoy lying in the sun. It's important not to ignore the practical alterations to living conditions made by people in response to temperature changes – they are as much a part of the human condition as are the physiological adjustments.

11. For example, imagine that the weather has suddenly turned very chilly. What changes do you think you'd make to your:

- clothing
- food and drink consumption
- level of physical activity.

12. Using these same three headings, list the changes you'd similarly make during a hot spell.

You might call these adjustments activities of living factors in temperature

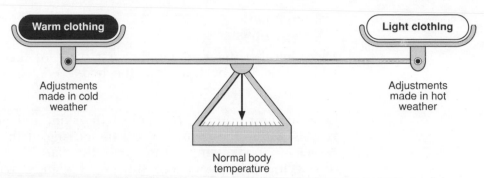

Fig. 1.9 Activities of living factors in temperature regulation. Complete your diagram by adding further activities of living.

regulation. I've tried to express some of these factors in diagram form as a way of aiding your understanding. I have begun such a diagram (Fig. 1.9) which you should try to complete.

13. What is the normal body temperature? You'll find this is a narrow range rather than a single figure. You should note how body temperature varies – within its normal limits – from early morning to evening to middle of the night.

14. What is the difference between the core temperature and the peripheral temperature of the body? What constitutes the core of the body with respect to its temperature, and what major organs are contained there? Is this core a fixed and unchangeable area, or does it vary? (See *Rutishauser 1994, Fig. 16.1.*)

Body temperature is the result of heat gained (from the environment, metabolic processes, food and drink, etc.) balanced against heat lost (to the environment, etc.). Since this Work Sheet is concerned with the skin, I'm going to restrict the subject to the way in which heat is gained and lost from the body via the skin. Further information about temperature regulation can be found in: *Rutishauser 1994, Ch. 16*; and, in diagram form, in *Mackenna & Callander 1990, pp. 44–45.*

15. Heat is gained by the body from its environment by:

   - conduction
   - convection
   - radiation.

Make notes on these three forms of heat gain, and try to express them in diagram form. To help you, here are three simple questions.

   a. What is the commonest source of radiant heat?
   b. When you hold your hands a short distance above a central heating 'radiator', which of the above three methods conveys most heat to you from the metal surface? (Hint: the word radiator is something of a misnomer.)
   c. On a winter's day you make yourself some hot coffee and, before drinking it, you wrap your hands around the mug. In which of the above three forms is heat conveyed to your hands? (Heat will also be gained by your body when you've drunk the coffee, but we aren't considering that here.)

16. Now we turn to ways in which the body loses heat. Heat is lost by:

   - conduction
   - convection

- radiation
- evaporation.

Make another diagram for the first three of these, demonstrating ways in which heat can be lost from the body. We'll consider evaporation a little later.

Look at your original diagram of the skin, and note the presence of blood vessels within the dermis. We'll be looking at the response in size of these vessels to body temperature, and the part they play in either losing or conserving heat. Note also the sweat glands, which we'll be considering later.

In warm weather, our bodies try to lose heat. Heat is lost from the periphery of the body, thus protecting the core temperature. Heat passes from a comparatively warm object to a comparatively cool object, for example by radiation. The body will lose more heat by convection if the skin is exposed to moving air, by removing clothing and standing in a breeze or in front of a fan.

17. Find out what happens to the blood vessels in the dermis in response to hot weather. How do they alter in size when heat needs to be lost from the body?

 COMMENTARY ON QUESTION 17

You may restrict your answer simply to describing the changes in size of lumen of the peripheral blood vessels; or, if you wish, you may include the nervous control of this activity. If the latter, you'll need to discover the whereabouts of the temperature regulating centre and the vasomotor centre, and which nerves convey the 'instructions' to the smooth muscle in the walls of the relevent blood vessels. (This material is also covered in Units 2 and 3.)

Remember: if heat is to be lost from the skin, it needs to be warmer than the surrounding air. Which vasomotor activity will warm the skin – vasodilation (the blood vessels getting bigger) or vasoconstriction (the blood vessels becoming smaller)?

18. Now show how the peripheral blood vessels will react to cold weather, when heat needs to be conserved. Will they dilate or constrict? (Which will divert blood from the periphery of the body to warm the core organs?)

Try to answer Questions 17 and 18 in diagram form as well as by making brief notes.

19. I've already drawn your attention to the sweat glands in the skin. Explain the meaning of the terms sensible and insensible when applied to loss of water by perspiration (sweat).

20. In hot weather, when the body needs to cool down, changes occur in the size of blood vessels in the dermis. You've already discovered whether they dilate or constrict. At the same time, sweat is secreted on to the surface of the skin. What happens to it?

COMMENTARY ON QUESTION 20

Sweat that stands in droplets won't cool the skin. This tends to happen in a very humid climate. The skin temperature only falls when, assisted by the change in blood vessel diameter, the sweat e_____s. What is this important word?

You may like to discover what nervous activity causes sweat glands to increase secretion. You'll find this described in *Ch. 16* of *Rutishauser 1994*.

21. Why do you think dry warmth (i.e. a Mediterranean coast climate) is said to be far more comfortable than the moist warmth we get in Britain?

22. In cold weather, we tend to turn up the heating in our homes and offices. Many elderly people living off their pensions cannot afford too drastic an increase in fuel bills. What advice about clothing might you give an elderly person in such a situation? Is there any advice you could also give about food and drink intake?

FURTHER STUDY

The advice about clothing and 'internal fuel' (food and drink) given to an elderly person during winter is practically the same as one might give youngsters setting out on a hill climbing expedition in Britain. If in Question 22, you simply gave the advice, 'put on warm clothing', this is insufficient. Why is it that several layers of light clothing (e.g. vest, woollen shirt, two light sweaters) keep you warmer than one very thick sweater?

Out on the hills, should the young climbers get wet through, they will also become very cold. Why is this? Why does dry clothing provide better insulation than wet clothing?

What is the wind chill factor? Why do we feel colder in a brisk wind than in still air? (I know this is common sense, but what's the physiological explanation?)

If one of the climbers becomes thoroughly chilled so that his core temperature falls, what is the name of this condition? What are the signs the other climbers might spot? What first-aid measures should they immediately put into practice?

23. Taking a patient's temperature is a very commonly performed nursing observation, and one that should be carried out accurately. You should read about taking patients' temperatures in: *Rutishauser 1994, Ch. 16, p. 298; Hinchliff & Montague 1988, pp. 571–572.*

What is the term for a patient's temperature that has risen to, say, 38.5°C? This can happen because of a wound infection or a urine infection. If this is the case, the patient may be given a course of antibiotics. What evidence would you look for that the antibiotics were having some effect? (Think about the basic nursing observations you'd be making. How would you evaluate the care you were giving?)

FURTHER STUDY

If you're interested in sports physiology and, in particular, temperature regulation during strenuous exercise, I'd recommend *Bursztyn 1990, Ch. 7*.
The cartoon-type diagrams are very helpful.

■ Bones, joints and muscles (guided study)

Time for completion	About 7 to 8 hours

Overall aim

To examine how the structure of bones and joints, and the contraction of muscles, contribute to the safe and efficient movement of the human body.

Introduction

The structure of the skeleton and muscles is studied not just by nurses and physiotherapists, but by artists. The drawings of nudes by Leonardo and Michelangelo reveal, beneath the subject's skin, a detailed knowledge of bones, joints and muscles, which is why these drawings look so life-like.

When nurses and physiotherapists work with patients, positioning them in bed or exercising weakened limbs, they use their understanding of joints and muscles in order to move the patients' limbs correctly.

Note, for example, how far back you can move your arm from the shoulder, or how far round you can turn your neck. Now imagine how painful it would be if someone were to move your arm or neck too far, or in the wrong direction. As with other areas of the body, knowledge of how joints and muscles function helps nurses care intelligently and appropriately for their patients. The correct positioning of a stroke patient's limbs can make all the difference in the effectiveness of his or her eventual rehabilitation.

Background questions

1. Distinguish between the following broad types of bones:

   • flat bones
   • long bones
   • irregular bones

   and give two examples of each.

2. Periosteum and marrow can be found in, for example, the femur. What do the two terms mean? To which of the above three types of bones does the femur belong?

3. Physiotherapists define the direction of movement of the limbs or joints they're working on by using terms such as abduction and adduction. Find out what these two terms mean. Imagine that you're lying in bed with your legs straight out in front of you. Draw diagrams demonstrating abduction and adduction of the legs.

4. Similarly, define the terms flexion and extension as they apply to the neck and spine. Imagine you're sitting at a desk for long periods. What sort of stretching exercises might you make to stop yourself stiffening? Can you express those movements in a simple diagram?

5. One of the ball and socket joints found in humans is the hip joint, which we'll be examining later. Can you think of another example of a ball and socket joint? Describe the movements made by the hip joint. What are the names of the large thigh muscles controlling movement at the hip joint? Can you think of a disease that commonly affects the hips, often in elderly and overweight people?

For some of the above questions see: *Rutishauser 1994, pp. 473–474, 480 and 484; Marieb 1992, pp. 230–232.*

Specific questions

SECTION 1: DEVELOPMENT, FUNCTIONS AND HEALING OF BONES
Time for completion: about 2 hours

Bones provide a rigid framework to which muscles are attached, so that the body can move. They also give protection to the softer organs beneath. Think of the rib cage, for example, helping to protect the heart and lungs from crush injury, and the skull protecting the brain.

Bones act too as a sort of reservoir for minerals such as calcium. Calcium can be taken from bones in some circumstances, and in others laid down again.

It may be hard to think of bone as a living and changing substance. Yet, like softer tissue such as muscle, it requires its own supply of nutrients. This means that fractured bone can usually heal. Bone starts to develop within the fetus, and continues to grow until the person is in his or her mid-20s. Conversely, especially in the elderly, bone can be weakened by the withdrawal of important minerals, until fractures can occur. Bone is very much living material.

1. Draw a diagram of a long bone, such as a femur, showing its shaft and, at either end, its epiphyses. You'll be adding to this, so make the drawing large.

2. Show on your diagram where the following are to be found:

   - periosteum
   - compact bone
   - cancellous bone
   - yellow bone marrow
   - red bone marrow.

   Make brief notes, using your preferred textbook, on each of these. (See, for example, *Rutishauser 1994, Ch. 28.*)

3. At either end of the long bone, the periosteum merges with a smooth layer of very hard tissue called articular cartilage. Add this to your diagram. What does the term articular mean? Why do you think this cartilage is so smooth and hard-wearing?

4. If you've ever fractured a bone, you will know how much it hurts. Whereabouts in bone are nerve fibres found? Note your answer, but for the sake of clarity it's perhaps best not to add nerve fibres to your diagram.

 ACTIVITY

If your College has a skeleton or a collection of bones, examine them and see if you can see the difference between compact and cancellous bone. Also, note how relatively lightweight bone is for its size and strength.

---

5. Make a series of small-scale diagrams showing how long bones develop from before birth to adulthood. Make accompanying notes, explaining how bone tissue is formed, and how bones lengthen as the body grows. Your notes should define the following:

   - osteogenesis (or ossification)
   - osteoblast
   - osteoclast
   - osteocyte.

   See: *Wilson 1990, Fig. 16.4; Hubbard & Mechan 1987, pp. 38–40 and Fig. 1.41;*

Rutishauser 1994, pp. 471–473; Marieb 1992, pp. 163–167 and Fig. 6.8; Hinchliff & Montague 1988, pp. 231, and 233–234.

6. Explain how blood is supplied to bones. You'll need to trace the arrival of blood via the arterial system, how it enters the periosteum (the outer covering of the bone) and how it is dispersed within the bony tissue itself.

 COMMENTARY ON QUESTION 6

Have a look at *Marieb 1992, Figs 6.2 and 6.4*, which clearly show how blood vessels enter bone via the periosteum, and how they are found deep within the bone. You should read about what is referred to as the matrix of bone – its internal organisation or structure – and the place of the following in that matrix:

- Haversian canals
- canaliculi
- lacunae.

(I'm sure you already know what a canal is (if not the Haversian version) but what do the terms canaliculus and lacuna mean? Have a look in the Glossary of Marieb 1992.)

You will discover that bone consists of a series of fine tubes, rather than completely solid material. Such a structure provides strength and comparative lightness, as well as allowing the passage of nutrients down the canals.

In the event of a fracture of a major bone, such as the femur, not only is there a lot of pain, but there is swelling of the surrounding tissues. On examining a victim's leg, even when the broken bone does not protrude through the skin, the thigh appears misshapen and swollen. Can you think why the thigh should appear so swollen? Hint: the answer may relate to Question 6 of this Guided Study.

---

7. Explain how bone heals following a fracture. You should use a series of small-scale diagrams to supplement your notes. What are the functions of osteoblasts and osteoclasts in this process? See: *Wilson 1990, p. 373; Rutishauser 1994, p. 473; Marieb 1992, pp. 171–172; Hinchliff & Montague 1988, pp. 237–240.*

 FURTHER STUDY

You may wish to add to your notes simple diagrams showing different types of fracture; for example compound, complicated, comminuted. What is the function of a splint, as applied by paramedics after a road accident? What is meant by 'reducing' a fracture? What part does a person's diet play in the successful healing of a fracture?

---

8. What effect does lack of exercise have on a person's bones? When someone is relatively immobile for a long period, what actions can nurses take to ensure that the patient's limbs are exercised as much as possible within, of course, medical restraints? Lack of exercise can seriously affect the structure of bones. If you follow up the references below, you'll discover that osteoporosis – the condition caused by lack of exercise – has other causes too, including prolonged weightlessness as in space flights. See: *Rutishauser 1994, Chs 13, 17, and 28; Hinchliff & Montague 1988, p. 237; Marieb 1992, p. 173; Guyton 1991, p. 881.*

Have a short break here.

SECTION 2: STRUCTURE AND MOVEMENT OF SYNOVIAL JOINTS
Time for completion: about 2 hours

You've already described some of the movements possible by certain joints, including a ball and socket joint like the hip. Remind yourself of the meaning of the terms abduction and adduction. Joints enable the body both to change the position of its limbs and to move as a whole. In this Section of the Guided Study we're going to concentrate on synovial joints, such as the hip and shoulder.

9. First, however, make brief notes on the other types of joint:

   - fibrous
   - cartilagenous

giving examples of each. How much movement is possible in these types of joints? Consider, for example, the healthy spinal (or intervertebral) joints.

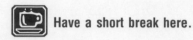 **COMMENTARY ON QUESTION 9**

See *Marieb 1992, pp. 223–226* and *Rutishauser 1994, Ch. 28* for this question.

Try to analyse anatomical terms you come across in your studies. For example, intervertebral tells you exactly where these cartilagenous joints can be found. Even the most frightening-looking terms can usually be broken down into more handleable bits.

Now we look at synovial joints, those many joints in the body (knee, hip, shoulder, etc.) where there is a fluid-filled cavity separating the bones involved.

10. Copy out a large version of Figure 1.10, large enough for you to add further details later. This figure is not meant to be of any specific joint in the body, but represents common features of synovial joints. The structures you need to add to Figure 1.10 are those forming the joint capsule:

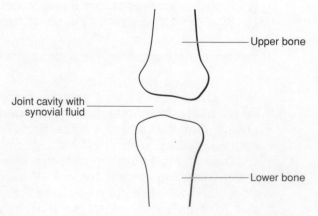

Fig. 1.10 Basic structure of a synovial joint. Add details to your diagram.

- the synovial membrane
- the hyaline or articular cartilage
- the fibrous capsule
- ligaments.

Make sure you understand the type of tissue forming ligaments and cartilage.

11. Describe the function of the synovial membrane. See: *Rutishauser 1994, p. 474; Marieb 1992, Fig. 8.3, p. 227.*

12. What is the appearance of hyaline cartilage, and what is its function?

 ACTIVITY

If you have the opportunity to carve or joint a chicken carcass, you'll be able to observe the hyaline cartilage on the ball joint of the bird's leg. Even cooking doesn't seem greatly to affect the hard shiny surface of this cartilage.

 FURTHER STUDY

What effect does a disease like arthritis have on certain of our joints? Although we're not studying pathology here, nor will we distinguish between the various forms of arthritis, find out how the smooth hyaline cartilage on a hip joint changes in arthritis. You'll be able better to understand the clinical features of the disease. Can you find a radiograph of a hip replacement? Look in a surgical nursing textbook. See how the surgeon's artificial joint (or prosthesis) mimics nature's original.

13. Explain how structures such as the bursa (pl. bursae) act as shock absorbers within very active joints such as the shoulder. Note the position of bursae in the shoulder joint. What other joints do you think contain their own version of shock absorbers? Think of how an athlete lands after a high jump. What parts of her body are jarred by landing badly? How does she attempt to protect these parts of the body? How do they protect themselves by absorbing the jolt?

14. The hip joint is one of the most freely moveable in the body. You've already noted movements such as abduction and adduction. Describe what other types of movement are possible in the hip, and draw simple diagrams to illustrate them.

15. Now compare these with the movements possible in the:

- neck
- knee
- wrist
- shoulder.

See: *Marieb 1992, Figs 8.6 and 8.7 (pp. 230–232); Rutishauser 1994, Figs 28.20 and 28.21.*

Now we look at how muscles and bones work together to produce movements.

16. Look at *Mackenna & Callander 1990, p. 281*, which shows the arm acting as a lever. What are the two large upper arm muscles at work here, and how are they attached to the bones involved? (You could also name the three main bones in the diagram.) See also *Rutishauser 1994, Fig. 28.18*.

17. Which muscle contracts when the lower arm is lifted (flexion)? What happens to the other muscle?

18. Which muscle contracts when the lower arm is pushed down (extension) as in the diagram in Mackenna & Callander? Again, what happens to the other muscle?

19. Try to distinguish between the terms agonist and antagonist as they apply to the muscles in the diagram. (Note that these terms are also used in pharmacology.)

Usually, we are well aware of the position of our limbs. We don't need to look around us to discover that we're standing upright or lying flat. We don't need to check to know that a hand is clenched, or that our feet are slightly apart. Somehow, the body 'knows' its own position.

 ACTIVITY

You can check this by carrying out some simple exercises. First, stand up, with your legs slightly apart and your hands by your sides. Close your eyes. Now touch your left knee with your right hand. Then touch your right elbow with your left hand. The odds are you made contact correctly both times. Finally, still with your eyes shut, touch the tip of your nose with a forefinger. The nose is a smaller target than either your knee or elbow, but it's likely you managed to do this accurately too. How does the body 'know' the relative position of its different parts?

20. Information about the position of limbs and joints, and tone of muscles, is passed to the brain from specialised sensory organs situated within muscles, tendons and joints. These sensory organs are called proprioceptors. *Marieb 1992, Fig. 13.12, p. 449*, shows these, and on *p. 424* there is an explanation of the term 'proprioceptor'. The diagram in *Mackenna & Callander 1990, p. 259*, shows the structures and sites of the different proprioceptors. One type, situated near or within joints, is called a Pacinian corpuscle. (Where else in the body are Pacinian corpuscles found?)

We'll return to the specialised functions of muscle spindles in Section 3 of this Guided Study. Meanwhile, make brief notes on the functions of each type of proprioceptor, perhaps utilising a simple diagram of your own. If you look at *Mackenna & Callander 1990, p. 260*, you'll see the various nerve pathways from the proprioceptors to the brain. The diagram here is somewhat complex, so just note that different levels of the nervous system give rise to different levels of awareness of the position of the body in space (as the textbooks put it). It's only when nerve impulses reach the cerebral cortex that the individual is fully aware of changes in the position of his limbs. Swift reactions to changes in position, as when we start to lose our balance, are instigated much lower down in the nervous system – see Unit 2 of this Workbook.

Once you've obtained a clear overview of this subject, read through

'Control and coordination' in *Ch. 28* of *Rutishauser 1994* for a more detailed discussion *(pp. 488–489)*.

 This is a good place for a fairly long break.

SECTION 3: CONTRACTION OF MUSCLES
Time for completion: about 3 hours

This is a complex subject, and my aim is to help you build up your knowledge step by step. We'll study the structure of muscles, the role of calcium in contraction, how certain muscles are stimulated to contract, and how contraction is controlled. Because there are close links with the nervous system, which is not studied until Unit 2, I'll be introducing relevant nervous system topics at a basic level, leaving their more detailed study until later. (This is one of the problems of trying to divide physiology into neat compartments, when in fact the different body systems are linked with each other.)

21. First of all, make brief notes on the three broad types of muscle:

    - skeletal
    - smooth
    - cardiac.

You should note differences in appearance of the three types of muscle, examples of where each type is found, and what sort of nerve supply each type has.

 COMMENTARY ON QUESTION 21

It's fairly obvious where cardiac and skeletal muscle might be found; but which organs contain smooth muscle? As to appearance: why is some muscle referred to as striped muscle? The terms voluntary and involuntary are also applied to muscle types. This may help you work out which nerves supply each broad group of muscles. You may at this moment be demonstrating the voluntary nerve supply to your arm and hand muscles, by turning over the pages of this book and by writing notes. But can you 'order' the muscle in the small intestine to contract and relax? If you cannot, it would suggest that the nerve supply to these muscles is involuntary.

Base your notes for Question 21 on the following short references: *Hubbard & Mechan 1987, pp. 40–41* (read just the introductory paragraphs to this section at the moment); *Rutishauser 1994, Ch. 22* (read the introduction, then the first few paragraphs on each of the three types of muscle); *Marieb 1992, pp. 246–247*.

22. Now let's look at a skeletal muscle, such as the biceps, and break it down into its functional units. First, describe the gross anatomy of the muscle, noting the arrangement of muscle fibres and connective tissue, and how the former are bundled together. How is the muscle fixed to bone? Make notes on:

    - tendon
    - epimysium
    - endomysium

- fasciae
- fasciculi.

See: *Hubbard & Mechan 1987, Fig. 1.43, p. 41; Marieb 1992, pp. 248–251 and Fig. 9.1.*

23. You'll have discovered that the primary bundles of muscle fibres (fasciculi) can be arranged in different patterns according to the site of the muscle and the work it is meant to do. What arrangement of fasciculi is found in the biceps? And in the pectoralis major? See *Marieb 1992, Fig. 9.3, and pp. 249–251.*

Now we examine the microscopic structure of skeletal muscle. First, let's discover why it is also referred to as striped muscle. The striped pattern found in skeletal muscle can be clearly seen in the microphotograph in *Marieb 1992, Fig. 9.4(a),* but we need to discover the significance of those stripes.

Note that cardiac muscle also has stripes. There are many similarities – and some differences – between skeletal, cardiac, and smooth muscle. Because skeletal muscle is probably the most familar to us (simply because we can feel these muscles under the skin, and see them change shape as they contract and relax) I'm going to concentrate at first on these striped, voluntary muscles.

24. Draw a diagram showing the relationship of a myofibril to a skeletal muscle fibre. Just as a muscle can be broken down into many individual muscle fibres, so a single fibre consists of many myofibrils. See: *Hubbard & Mechan 1987, Fig. 1.44; Mackenna & Callander 1990, p. 283; Marieb 1992, Fig. 9.4.*

On the myofibril that you've drawn note the alternating bands of light and dark. These are the stripes of striped muscle.

25. These bands form a regular recurring pattern. On your diagram map out one sarcomere, that is, one whole pattern of light and dark which is repeated throughout the length of the myofibril. Figure 1.11 demonstrates the relationship of each part of a muscle to the whole, and also the sarcomere, which is the functional or contractile unit of a muscle. As we study the sarcomere in more

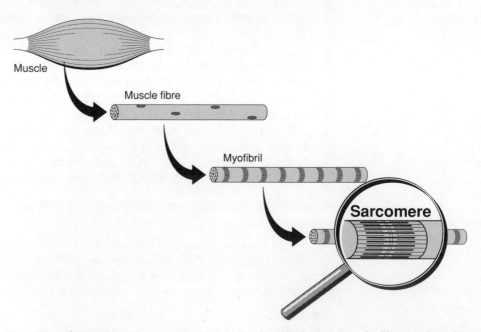

Fig. 1.11 The relationship of one sarcomere to its myofibril, to a muscle fibre, and to the striped muscle as a whole.

detail, remember this figure in order to retain an overview of what happens when skeletal muscle contracts.

26. The different patterns of light and dark are labelled to help describe what happens when muscle contracts. Label the different regions in your diagram:

    - Z lines
    - I bands
    - A band
    - H zone (this incorporates the A band).

This is a good point to take stock. We've discovered how muscle (in this instance skeletal muscle) can be broken down into smaller and smaller units, until we arrive at the bandings on a single myofibril. Physiologists noticed that the pattern of light and dark bands forming the sarcomere changes during contraction. So events at the sarcomere will be our next item of study.

 If you need a short break, this would be a suitable place.

The dark and light patterns that make up the stripes of both skeletal and cardiac muscle consist of arrangements of thick and thin filaments.

27. Describe the molecular structure of myosin, which makes up the thick filaments. Myosin molecules are arranged along the thick filaments in a certain way – describe this, and perhaps draw a diagram to help your notes.

28. Do the same – in notes and diagram – for the actin molecules, which make up the thin filaments. Note the role of tropomyosin and troponin in the structure of the thin filaments. Whereabouts are there calcium binding sites? (We'll see later how important calcium is for muscle contraction.)

29. You already know that myofibrils are bundled together, held in place by an outer 'wrapping' called the sarcolemma. In relation to this sarcolemma, what are the sarcoplasmic reticulum and T tubules? Show them in diagram form, with brief notes to explain their appearance and function.

30. In your notes you should also explain the presence of many mitochondria. If you completed the earlier Work Sheet on the cells of the body, you should be able to suggest their function.

For Questions 27 to 30 see: *Rutishauser 1994, pp. 384–385; Hubbard & Mechan 1987, pp. 41–46; Jennett 1989, pp. 90–93; Marieb 1992, pp. 251–254; Guyton 1991, pp. 67–68.*

31. Now we need to describe the mechanism of contraction – what it is that makes the filaments slide across each other (rather like your open fingers interlocking as you bring your hands together). You should discover the relevance of the myosin molecules with their 'twin heads' arranged in a regular pattern along the thick filaments like pairs of oars along a rowing boat. You should read about the role of calcium in this sliding filament process. (We'll leave the question of how this process is powered until the next question.)

    For a good summary of the mechanism of contraction and the role of calcium, see *Rutishauser 1994, pp. 385–386.* Then, in increasing order of detail, refer to: *Hubbard & Mechan 1987, pp. 46–50; Jennet 1989, pp. 92–93; Marieb 1992, pp. 254–258.*

Then, if you're sure you are clear about the basic mechanism of muscle contraction, try *Guyton 1991, pp. 69–72*. (Don't be frightened of books of this size and complexity; never be afraid to dip into them in order to gain additional information. Always, however, be certain that you understand the fundamentals first, or you may be confused rather than enlightened.)

32. Now read about, and make notes on, how our muscles gain the power they need to contract. Athletes often eat specially chosen foods prior to a race, depending on whether it's a long run like a marathon, or a short burst of energy as in a 100 metre race. Bear in mind, though, that one major muscle of the body requires energy all the time, whether we are active or fast asleep. Which muscle?

    Rutishauser 1994, Ch. 17, provides information about how blood flow to skeletal muscles alters during exercise, how ATP is formed and broken down, and how waste products are formed during muscle activity. But also see *Ch. 22* for a discussion of 'slow' and 'fast' skeletal muscles, and how our muscles develop from childhood to adulthood.

    You could also read *Marieb 1992, pp. 266–269*. Can you distinguish between aerobic and anaerobic muscle metabolism? Why do our muscles ache when they have exercised a lot? What is the oxygen debt? (Try running up several flights of stairs – you may be breathing heavily for some time afterwards. Why is this?)

 FURTHER READING

What happens to our muscles if we are confined to bed for a long period? (Bear in mind that with some fractures, a person can be in bed for about 12 weeks.) Are the changes you discover inevitable, or can appropriate nursing care help to prevent them? Don't assume, by the way, that when a person is paralysed by a fractured spine, he is necessarily inactive. Have you seen wheelchair basketball on television, or wheelchair athletes in the London Marathon?

Skeletal muscle contraction is triggered by nerve impulses. The nature of nerve impulses will be studied later, but, in preparation for this, you should find out how motor nerves (nerves that stimulate movement) join their appropriate skeletal muscle.

33. So, find out the meaning of the term motor unit, and draw a diagram demonstrating a single motor unit. See: *Rutishauser 1994, p. 396; Marieb 1992, Fig. 9.13, p. 263*.

    You should note how a motor nerve (also called a motoneurone) divides into branches each of which ends in a motor end-plate. Each end-plate innervates one nerve fibre.

    So does one motor nerve supply just one muscle fibre, or many muscle fibres? If the latter, does this vary in number? (See the Rutishauser reference above.)

34. Now describe, in words and a diagram, a neuromuscular junction – that is, the point where nerve and muscle fibre meet. Do they actually touch? See: *Rutishauser 1994, Fig. 22.18; Marieb 1992, Fig. 9.10.*

COMMENTARY ON QUESTION 34

A chemical is produced from the nerve ending at the neuromuscular junction when a nerve impulse arrives. You should discover the name of this chemical and, in simple terms, its function. What does this chemical do when it meets the receptors on the muscle fibre, and what happens to it afterwards?

Later, in Unit 2, you'll discover that events at the neuromuscular junction are fairly similar to those at a synapse (join) between two nerves. We'll leave discussion of the detail of electrical activity to Unit 2. Here, be content simply to clarify the structure of the neuromuscular junction, and the succession of events when a nerve impulse arrives there.

35. Find out the meaning of the term recruitment, in respect of the activation of motor units.

36. Finally, with regard to skeletal muscles, find out what the term muscle tone means. See *Rutishauser 1994, p. 462.* (Earlier in this Guided Study, there was a brief reference to muscle spindles. Here you'll find rather more detail about the role and innervation of these important structures in maintaining muscle tone.)

For much of this Guided Study, we've concentrated on skeletal muscle, with only the occasional reference to smooth and cardiac muscle. They have many similarities with skeletal muscle, and some differences. You'll discover what these similarities and differences are by reading *Ch. 22* of *Rutishauser 1994*. After an introduction which provides a general overview of muscle contraction, the author deals first with smooth muscle – its structure, its innervation, and how muscle contraction is triggered. Then the same sequence of description is applied to cardiac muscle. Finally, skeletal muscle is covered using exactly the same format.

37. You've already read parts of this chapter in your study of skeletal muscle. My suggestion is that you read, first, the whole of the section of Ch. 22 dealing with skeletal muscle. This should be useful in fixing your knowledge, and reassuring you that you understand the notes you've already made.

38. Now go back to the section on smooth muscle, read through it, and make notes. You should understand what it is that triggers contraction of smooth muscle and how this differs from skeletal muscle stimulation. How are adjacent muscle cells excited in smooth muscle? What happens when, for example, the gut is stretched by the presence of food? What effect does this have on the smooth muscle in the gut walls?

39. Finally, read through the section on cardiac muscle, and make notes. Though cardiac muscle is striped, there are important differences between cardiac and skeletal muscle. You'll discover that cardiac muscle cells are closely linked together – but for what purpose? What is it that triggers a muscle contraction? What is the purpose of the autonomic nerves supplying the heart? (You will study the autonomic nervous system in more detail in Unit 2.) What is the all-important conducting system of the heart muscle, and what is the meaning of the term pacemaker? How is it that the heart can keep going, from before we're born to the day we die? Where does its energy come from?

Your notes and diagram(s) on cardiac muscle will be very useful when you come to Unit 3 of this Workbook.

You have probably amassed many pages of notes, perhaps too many to look through now before concluding this Guided Study. I'd suggest, however, that before you close your books, you take a look at the diagrams that you've drawn. Are they properly labelled, and is their purpose clear? Remember that a good diagram can summarise a lot of words, and can be a useful means of revising a subject.

 Now have a well-deserved break.

■ Skin, bones and joints (check quiz)

1. Explain how the skin helps to protect a person from infection. On your way to work, you trip on a loose paving slab and fall, grazing your hand. What implications might this have for your personal safety while nursing on a surgical ward?

2. How does the skin help to maintain a person's temperature within normal limits? What nursing care, based on your knowledge of physiology, would help lower the temperature of a pyrexial patient?

   What advice might a district nurse give to an elderly lady living at home, to help her keep warm during the winter? Again, describe the physiological principles involved.

3. Explain how bones heal following a fracture.

   Discuss with a group of your student colleagues the purpose of immobilising a fractured limb:

   - as first aid, for example immediately following a climbing accident
   - after the fracture has been 'reduced' in Accident Service.

   (Concentrate on the principles of immobilising an affected limb.)

4. State the similarities and differences between smooth, cardiac, and skeletal muscle. What part does calcium play in muscle contraction?

5. Describe the appearance and function of hyaline cartilage, for example in the hip joint. What is the function of the synovial membrane?

6. The physiotherapist on your ward asks that your patient should carry out:

   - abduction and adduction exercises to the arms
   - flexion and extension of the spine.

   Within a group of your colleagues, demonstrate these exercises and name the joints that are involved.

   (You could demonstrate these exercises yourself or, perhaps rather trickier, by using one of your colleagues as a 'patient' and moving his or her joints. If the latter, be very careful not to force the 'patient's' joints beyond their normal range of movements.)

Control and communication

Contents

- **An overview of the nervous system**
  Work sheet (5 to 6 hours)

- **Further exploration of the nervous system**
  Guided study (2 days)

- **Pain**
  Suggested reading

- **The nervous system**
  Check quiz

- **Introduction to endocrines — thyroid and parathyroid glands**
  Work sheet (2 to 3 hours)

- **Further exploration of the endocrine system**
  Guided study (8 to 10 hours)

- **The endocrine system**
  Check quiz

- **The special senses — hearing and sight**
  Suggested reading

UNIT 2

■ An overview of the nervous system (work sheet)

Time for completion About 5 to 6 hours, or one whole 'college day'. Note that this is a little longer than the time usually devoted to a Work Sheet.

Overall aim To achieve a basic understanding of the structure and function of sensory receptors, sensory and motor neurones, and sensory and motor areas of the cerebrum.

Introduction The nervous system may seem rather frightening for students to tackle because it has so many different parts. It is easy to become so involved with all these 'bits' that we lose sight of the function of the whole. Neither this Work Sheet nor the following Guided Study can hope to cover every detail, but they aim to examine certain aspects within the context of the whole system. As you study each part, try to remember where and how it fits in the overall scheme of the nervous system, and how it contributes to the functioning of the human body.

A properly functioning nervous system allows us to respond both to danger and to the beauty of a symphony or a painting. It allows us to listen to what others have to tell us, and to reply. It allows us to fight, and to hug.

Many functions of the nervous system occur below our level of consciousness – we can't 'instruct' our heart rate to increase or decrease or our arterioles to dilate or constrict.

Such automatic adjustments are brought about by the activity of the autonomic nervous system (sometimes in conjunction with the release of endocrines) which is covered in the next Guided Study. This Work Sheet concentrates on the voluntary control of body activities.

Background questions 1. It is helpful to obtain an overview of the nervous system so that its many complex facets may be seen in context. Not all physiology texts provide such an overview, so you may find *Rutishauser 1994, pp. 351–353*, particularly helpful. You should find the link between physiology and psychology, and the explanation of the nervous system's difficult terminology, a helpful basis for further study.

2. Copy and label the diagram of a neurone (Fig. 2.1) in order to learn its different parts. In your reading, you will come across terms such as sensory neurone and motor neurone, myelinated and non-myelinated neurones, but don't concern yourself with these distinctions at the moment. Your diagram is of a basic neurone, millions of which form the various structures of the

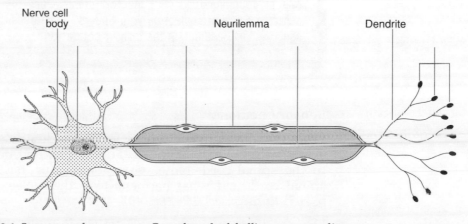

Fig. 2.1 Structure of a neurone. Complete the labelling on your diagram.

nervous system. Refer to: *Wilson 1990 p. 241, Fig. 12.2; Rutishauser 1994, Fig. 21.1.*

Specific questions

In the course of this Work Sheet we'll be following a 'message' from its source in sensory receptors in the skin, along nerves to the brain, looking at how the brain deals with this sensory message, and how it responds by sending instructions along motor nerves to certain voluntary muscles. For an outline of this nerve network, see *Rutishauser 1994, Fig. 3.12.*

RECEIVING MESSAGES: THE SENSORY PATHWAYS
Time for completion: about 1½ hours

Sensory receptors

1. Sensory nerve endings can be found throughout the skin (and also in organs such as the intestines, lungs and blood vessels). First, list the types of sensation experienced by different stimulation of the skin. You will probably easily think of pain and light touch, but there are several more.

2. Draw a diagram showing the different types of sensory receptors found in the skin. Give the name of each type of receptor, and state the stimulus (i.e. touch, temperature, pressure) that excites it. See: *Rutishauser 1994, p. 403; Hubbard & Mechan 1987, pp. 223–226; Guyton 1984, pp. 144–145; Mackenna & Callander 1990, p. 261; Marieb 1992, p. 424 and Table 13.1 on pp. 426–427.*

Pay particular attention to the receptors registering light touch, since we'll be following the 'messages' from these receptors up to the brain.

 ACTIVITY

Sensory receptors registering light touch are more numerous in some areas of the skin than others. Demonstrate this with the following experiment using the twin points of a compass.

Get a friend to prod the skin gently in the small of your back with the compass points (being careful not to puncture the skin!). The points should begin at about 1 cm apart. How wide apart are the points when you can distinguish two separate pricks?

Now, with your eyes closed, get your colleague to repeat the process, this time prodding one of your fingertips with the two points close together. Can you distinguish two separate points of contact on your fingers? What does this tell you about the number of sensory nerve endings in the skin of both areas?

Sensory neurones

3. From diagrams in your chosen textbook identify sensory nerve endings in the skin that register light touch. Follow just one neurone from these receptors to the spinal cord. Note whereabouts this 1st sensory neurone enters the spinal cord, and what happens to it there. See: *Mackenna & Callander 1990, p. 265; Rutishauser 1994, pp. 407–409.*

4. Draw a diagram showing this 1st sensory neurone from its nerve endings to the spinal cord.

COMMENTARY ON QUESTION 4

Ensure your diagram shows the position of the 1st sensory neurone cell body outside the spinal cord. Demonstrate on a cross-section of the cord where the sensory neurone enters the cord – through the anterior or posterior horn – and whether it continues up the cord or links (synapses) with a 2nd sensory neurone. (We'll be looking at synapses in detail in the following Guided Study.) See: *Mackenna & Callander 1990 p. 265; Marieb 1992 p. 413; Rutishauser 1994, Fig. 23.7.*

At the moment don't worry about long terms you come across in textbooks, like spinothalamic tract. We're concentrating here on major divisions of the nervous system, such as the spinal cord and the brain. Once you have a sound grasp of the structure and course of sensory nerves, you can then extend your knowledge to include more complex terms.

5. Find out where the 1st sensory neurone you've been following links (synapses) with the 2nd sensory neurone. At what point does the sensory pathway cross the midline of the body to the opposite side?

6. Where does the 2nd sensory neurone synapse with the 3rd?

7. Figure 2.2 shows, in highly diagrammatic form, the sensory pathway from a touch receptor to the brain. Check your own more detailed diagrams with it. Now add to Figure 2.2 another sensory pathway for either temperature or pain. Note the different positions of synapses between 1st, 2nd and 3rd neurones for these new pathways.

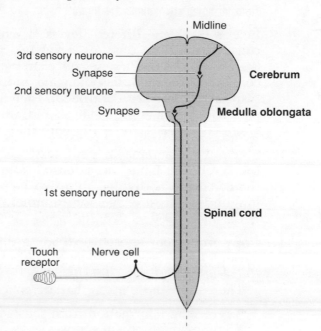

Fig. 2.2 Sensory pathway for light touch (diagrammatic form).

8. Note also from your textbook(s) how sensory nerves are bundled into different parts of the spinal cord according to where they enter the cord and their destination. (See *Rutishauser 1994, p. 356.*)

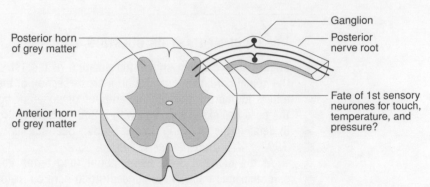

Ganglion

Posterior horn
of grey matter

Posterior
nerve root

Anterior horn
of grey matter

Fate of 1st sensory
neurones for touch,
temperature, and
pressure?

Fig. 2.3 Slice of spinal cord showing posterior (sensory) nerve root and ganglion (on left side only).

Add to the diagram of a slice of the spinal cord (Fig. 2.3) how 1st sensory neurones for touch, temperature and pressure enter the cord (perhaps using different colours to distinguish them), then ascend to their synapses with the appropriate 2nd sensory neurones.

9. What is it that causes the difference in colour of the grey matter and white matter of the spinal cord?

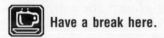 **Have a break here.**

INTERPRETING THE MESSAGE: THE CEREBRUM
Time for completion: about 1½ hours

Just as there are different levels at which our nervous systems operate (e.g. conscious and unconscious levels) so there are different structural levels to the brain. The brain stem, for example, is a continuation of the spinal cord, containing the medulla oblongata and the pons. Here body temperature and respiratory rate are controlled. Then there is the cerebellum, concerned in part with maintaining balance; and, finally, the cerebrum. In this Work Sheet we're concerned principally with the functions of the cerebrum.

10. Study the diagrams in your textbooks that show the outer appearance of the cerebrum. Note the extensive folding of the cerebral surface, and how the cerebrum is divided into two hemispheres – left and right – by a deep longitudinal fissure. See, for example: *Wilson 1990, Fig. 12.12, p. 248; Mackenna & Callander 1990, p. 222; Rutishauser 1994, Fig. 20.1, p. 353.*

11. Copy from your chosen textbook a diagram of the cerebrum, and label the various lobes. See: *Mackenna & Callander 1990, p. 222; Wilson 1990, Fig. 12.13; Rutishauser 1994, Fig. 20.11B.*

    Then show those areas that are concerned with:

    - receiving information from sensory nerves
    - organising motor responses
    - vision
    - hearing
    - speech.

You may find *Hubbard & Mechan 1987, pp. 235–243,* particularly useful here. Note the two diagrams of the homunculus, both motor and sensory. (*Rutishauser*

1994, pp. 409 and 466, has similar diagrams.) What does homunculus mean? What is the significance of those two diagrams?

12. Check with *Rutishauser 1994, Ch. 20*, that you understand anatomical terms, such as sulcus (sulci) and gyrus (gyri), that appear on diagrams in your textbooks. Note the position of the lateral sulcus, central sulcus, and longitudinal fissure. Do you understand the geographical terms longitudinal, lateral and central?

13. Copy out a diagram showing a cross-section of the cerebrum. (See, for example: *Rutishauser 1994, Fig. 20.3; Wilson 1990, Fig. 12.14*.) Show on your diagram the position of grey matter in the cerebral cortex. Note that whereas the grey matter of the spinal cord is in the middle of the cord, in the brain it surrounds the white matter. Check that you know the significance of the term grey matter.

14. Find in your textbook diagram a structure called the corpus callosum, and discover its function. Add it to your own cross-section of the cerebrum.

15. Similarly, add the thalamus and basal ganglia to your diagram, and make brief notes on their functions. See: *Rutishauser 1994, Fig. 20.11A; Mackenna & Callander 1990, pp. 222–225; Hubbard & Mechan 1987, pp. 241–242*.

16. Finally, add a sensory pathway from a touch, temperature, or pressure receptor. Note how, for example, the 3rd sensory neurones of temperature or touch pathways begin in the thalamus where they synapse with their 2nd sensory neurones. See: *Mackenna & Callander 1990, pp. 264–265; Marieb 1992, Fig. 12.25, p. 413*.

17. Make notes on the reticular activating system and its relevance to sensory input to the brain from the thalamus. See: *Mackenna & Callander 1990, pp. 223 and 225; Rutishauser 1994, pp. 410 and 522; Guyton 1984, pp. 203, 210 and 211; Marieb 1992, pp. 401–402*.

18. Read about, and make notes on, the limbic system, said to be the oldest part (in developmental terms) of the cerebral cortex. How are pain and rage, for example, perceived here, and how are such stimuli passed on to the cerebral cortex itself? See: *Marieb 1992, pp. 400–401; Rutishauser 1994, p. 525*.

19. Summarise the successive levels at which a person can become aware of a certain stimulus. See *Rutishauser 1994, Fig. 23.6 and pp. 409 and 423*.

20. What part of the cerebral cortex is concerned with the memory of sensations? Consider that, when you see or pick up a pencil, you know immediately what it is and how to use it. You have a visual and a tactile memory of the object that prevents you having to find out what it is each time you see or handle it. Our brains don't just memorise facts like dates or formulae; they hold a memory of the shape and feel and function of everyday objects. See *Rutishauser 1994, p. 512*. Try the following activity.

 ACTIVITY

This simple experiment demonstrates how well our brains memorise through senses other than vision and hearing. Have a friend collect a small group of everyday objects — pen, eraser, pencil, envelope, saucer etc. — out of your sight. Then with your eyes shut, try to identify each object by touch. You can extend the experiment by using various drinks, to be identified by only smell and taste.

Just as you can hum a tune you know well, so a pianist can play a piece of music from memory: not the visual memory of the music score, but the fingers 'knowing' how to move about the keyboard. This memory comes about by frequent practice, as any musician will tell you.

Before you finish this part of the Work Sheet, have a quick check through your notes and, especially, the diagrams you've drawn.

 Reward yourself with a longer break here. So far the Work Sheet may well have taken you the whole of a morning.

GIVING ORDERS: CEREBRUM TO MOTOR NEURONE
Time for completion: about 1½ hours

21. Find out which part of the cerebral cortex is concerned with motor responses. Show this on another diagram of the cerebrum. See: *Guyton 1984, pp. 182–184; Marieb 1992, pp. 385–386; Rutishauser 1994, Fig. 27.10.*

22. Check your earlier notes on the position and function of the basal ganglia. See *Rutishauser 1994, Ch. 27 and Fig. 27.12.*

23. Similarly, write a brief summary of the function of the cerebellum. (We'll be considering the cerebellum in more detail in the following Guided Study.) See: *Rutishauser 1994, pp. 467–469; Guyton 1984, pp. 184–187.*

24. Trace a motor pathway from the cerebral cortex to the voluntary muscle it innervates. Show where the upper motor neurone synapses with the lower motor neurone. Show on your diagram where the motor pathway crosses the midline to the opposite side of the body. See: *Mackenna & Callander 1990, p. 269; Rutishauser 1994, Fig. 27.1 and p. 464.*

25. Make notes on the motor unit – the joining of motor nerve and muscle fibres. See: *Mackenna & Callander 1990, p. 270; Rutishauser 1994, p. 461.*

 FURTHER STUDY

The motor response described above is highly simplified. When a muscle contracts, there is likely to be input from more than one motor nerve. When we move a limb, or turn our head, many parts of the brain, and many nerve fibres, are involved. These multi-neuronal motor pathways are described by *Mackenna & Callander 1990, p. 272*. Work steadily through this complex diagram, tracing the inputs from each of the areas of the brain shown there.

RESPONDING QUICKLY: THE WITHDRAWAL REFLEX
Time for completion: about 1 hour

What happens if you accidentally put your foot into a bath filled with water that's too hot? (You'll appreciate that I can't suggest you carry out an Activity to demonstrate this.) Obviously the foot is rapidly pulled out of the water – but is it a conscious movement? Do you think, 'My word, that water's hot, I'd better remove my foot, '(or words to that effect)?

If you think back to the occasions when this has actually happened to you, you'll probably realise that the following chain of events occurs:

- your foot enters the hot water
- there is a vague feeling of pain or discomfort and the foot is rapidly withdrawn
- about 1 second later the sensation of pain 'swells up' in your consciousness, and its source is pinpointed to your foot
- you may cry out because of the pain.

This is an example of a withdrawal reflex. The rest of this Work Sheet will be concerned with a simplified version of such a reflex, though I'll be pointing out texts which explain the complexity of what actually happens. For an overview of reflex action, see *Rutishauser 1994, Fig. 3.13*.

26. It's important to grasp that the withdrawal reflex is organised at the level of the spinal cord, not the brain (though information about sensations received and subsequent events is passed up to the brain). Look at one or more of the following diagrams: *Wilson 1990, Fig. 12.24; Rutishauser 1994, Fig. 27.3; Marieb 1992, Fig. 13.11.*

     Copy from these diagrams two of your own, one showing a two-neurone reflex arc, the other showing a three-neurone arc. (The middle neurone is called an interneurone or connector neurone.)

27. Add to your diagrams the following terms:

    - afferent neurone
    - efferent neurone.

I've avoided these terms so far because they are confusingly alike, and difficult to remember. One of them applies to the sensory neurone, the other to the motor neurone. See *Mackenna & Callander 1990, p. 231*, where these two terms are used. The diagram on *p. 233* usefully suggests how many other parts of the body may be involved in a reflex action. Imagine burning your hand on a hot plate. It's not just the hand that moves away. Your arm and shoulder move too; your head may jerk back, your chest expands as you take a sharp breath in, and your lips and tongue may form some appropriate verbal response.

28. Draw a diagram illustrating a stretch reflex, for example the knee jerk reflex often tested by doctors. See: *Mackenna & Callander 1990, p. 230; Marieb 1992, Fig. 13.13; Rutishauser 1994, Fig. 27.7.*

     How many neurones are involved in the reflex arc itself here – two or three?

29. Read about, and make notes on, how reflexes can be inhibited. An example commonly given is that, if you pick up a plate that is both very hot and very expensive, you can inhibit the initial reflex which would have caused you to drop the plate. Have another look at *Rutishauser 1994, Fig. 3.13B*. Where does this inhibition come from? See *Rutishauser 1994, p. 463*.

FURTHER STUDY

Read about muscle tone, and the control of delicate movements. You may have seen patients who have either flaccid muscles, or muscles that are very tensed (spastic). You may have noticed how some patients are unable to make well-controlled movements – they reach out for a glass of water and either miss it completely or send it flying. (Perhaps you've noticed – even experienced – the strange, exaggerated movements made by someone who is drunk.) See: *Rutishauser 1994, pp. 466–468; Marieb 1992, pp. 447–453; Guyton 1984, pp. 164–169.*

You may like to round off this Work Sheet by reading *Rutishauser 1994, pp. 511–513 and 517–519*, which will help to bring together much of what you've been studying in the context of the human person.

■ Further exploration of the nervous system (guided study)

Time for completion	Allow at least 2 'college days', with 5 to 6 hours' study per day

Overall aim

To build on the student's present knowledge by studying detailed physiology of both the voluntary and autonomic nervous systems.

Introduction

In the preceding Work Sheet we saw how a 'message' or nerve impulse travelled from sensory receptors in the skin to the brain, and how the consequent 'command' travelled from the brain to the voluntary muscles. We saw how another series of events could happen with extra speed in a spinal reflex, in order to withdraw parts of the body from a source of danger, such as a hot object.

In this Guided Study we look in more detail at nerve impulses – what they are, what causes them, how they travel from one part of the body to another. We also look at the nerve messages that occur below our level of consciousness, so that the body can function, as it were, on automatic pilot. We also look at the blood–brain barrier which has important clinical implications in, for example, the treatment of meningitis.

I repeat the warning given earlier, which is that by getting involved in intricate details it's all too easy to lose sight of how the nervous system functions overall. Especially, we can forget how physiology and psychology are intimately linked. Action potentials and neurotransmitters play a central part not just in the physiology of the nervous system, but in memory and learning, in mood, emotion and mental illness. Knowledge about nerve impulse transmission and synapses, for instance, assists our understanding of how certain drugs, such as antidepressants and alcohol, work.

Background questions

(Take about 1 hour for this introductory section.)

1. Revise your understanding of the structure of the neurone, distinguishing between the axon and dendrite.

2. Distinguish in both structure and function between unipolar, bipolar and multipolar neurones, and between sensory and motor neurones. See: *Marieb 1992, pp. 346–350; Rutishauser 1994, Fig. 21.1.*

3. Read about the different structure and appearance of myelinated and non-myelinated nerve fibres. (We'll see later how the presence of a myelin sheath affects the speed of nerve impulses.) See, for example, *Rutishauser 1994, p. 367.*

4. Read one or more of the following passages which show the links between the physiology of the nervous system and aspects of psychology: *Rutishauser 1994, Ch. 20*, the introduction and 'Understanding the nervous system' (you may have read this at the beginning of the previous Work Sheet but it will stand revisiting); *Rutishauser 1994, Ch. 33, pp. 543–544; Guyton 1991, pp. 478–480*, including 'Storage of information – memory'. You might also try *pp. 643–647*, 'Thoughts, consciousness, and memory'; *Marieb 1992, pp. 489–493*, 'Consciousness', 'Memory' and 'Language'.

Specific questions

SECTION 1: MEMBRANE POTENTIAL AND ACTION POTENTIAL
Time for completion: between 2 and 3 hours for this Section (e.g. a whole morning or afternoon)

Membrane potential

1. Like all cells, nerve cells have a fluid-like (or rather gel-like) interior and are

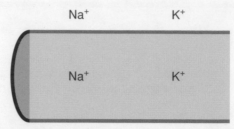

Fig. 2.4 Section of a nerve cell showing intra- and extracellular potassium and sodium.

bathed in fluid. Both fluids contain electrolytes. Figure 2.4 shows a small section of an axon, and the chemical symbols for two of these electrolytes, potassium and sodium. Note that they are both positively charged. (Does that make them cations or anions?) Find out the concentration of potassium and sodium, both intra- and extracellularly, and add them to your own copy of Figure 2.4. See: *Mackenna & Callander 1990, p. 56; Rutishauser 1994, pp. 29 and 369; Hubbard & Mechan 1987, pp. 56–57, including Figs 1.65–1.68; Marieb 1992, pp. 350–352.*

These references will serve for Questions 1 to 5.

2. What forces maintain these sodium and potassium levels? First there is the sodium–potassium pump, which moves both these ions across the membrane in particular directions. Add arrows to your diagram showing in which direction the pump moves (a) potassium ions, and (b) sodium ions.

3. Then there is movement of ions by diffusion. From the concentrations of both intra- and extracellular sodium and potassium, you'll realise that there are concentration gradients for both these ions across the membrane. Add arrows to your diagram showing the movement of potassium and sodium by diffusion.

4. However, the membrane is usually much more permeable to potassium ions than to sodium ions, and this is represented in Figure 2.5. Read about the balance between movement of ions by the sodium–potassium pump and movement by diffusion. What effect does the outward movement of potassium ions have on the electrical charge on the outside of the membrane compared with that on the inside?

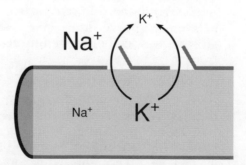

Fig. 2.5 Diffusion of K^+ through membrane pores.

5. What is the resting potential of the nerve fibre membrane, and how do physiologists measure it? Draw a diagram representing the resting potential of a nerve fibre, based on Figure 2.6.

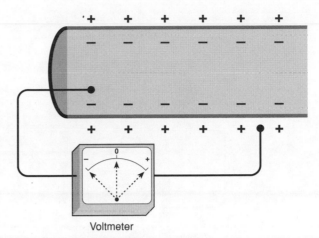

Voltmeter

Fig. 2.6 Resting potential of a nerve fibre. On your diagram, fill in the arrow that shows the appropriate voltage.

 COMMENTARY ON QUESTIONS 1 TO 5

After reading about this complex subject, you may find it helpful to write out a summary of factors in the creation of a resting potential. Try completing the following sentences:

- The sodium pump moves Na^+ ...
- Because the membrane is impermeable to Na^+ very little Na^+ diffuses ...
- The potassium pump moves K^+ ...
- Unlike Na^+, K^+ can diffuse ...
- The end result is a ... charge inside the membrane relative to the outside. This charge is said to measure ...

Note that texts give slightly different voltages for large and small nerve fibres.

Action potential

Differences in sodium and potassium levels are present across the membranes of all cells in the body, including nerves and muscle fibres. But stimulation of a nerve membrane can cause a chain of events – the action potential – that leads to an electrical current (nerve impulse) passing from one end of the nerve fibre to the other. This chain of events is based on very rapid, but transient, changes in membrane permeability.

6. One form of stimulation that can cause an action potential is an electric shock applied to a nerve (as in a laboratory experiment). What are the other forms of stimuli? (Think about the types of sensory receptors, not just in the skin but throughout the body.)

7. The stimulation – of whatever form – causes a sudden increase in membrane permeability to sodium, so that sodium ions rush inside the membrane. (Why inside?) This movement is shown on the right of Figure 2.7 (p. 52). Add the correct voltage for the action potential, caused by the movement of sodium ions.

8. The sudden permeability of the membrane to sodium is very transient, lasting only a few milliseconds. Immediately afterwards, the membrane reverts to its

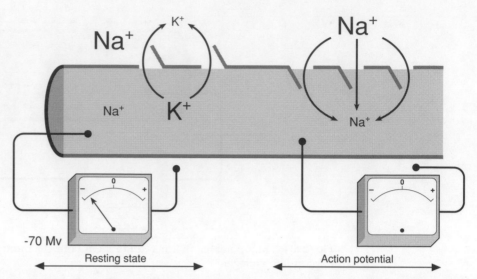

Fig. 2.7 Movement of Na⁺ to bring about the action potential. Add the appropriate voltage to your diagram.

usual state of being relatively impermeable to sodium. What happens now to the electrical charge across the membrane? How is this change brought about? (You'll need to refer to the movement of both sodium and potassium ions.)

9. The change in membrane permeability is brought about by voltage gating. Read about this in *Rutishauser 1994, Ch. 21*. Also define the terms:

   - threshold,
   - depolarisation,
   - repolarisation, and
   - all-or-nothing response.

Propagation of the action potential

The brief but explosive change in membrane permeability – the action potential – isn't confined to the part of the membrane that received the stimulus. If a stimulus is strong enough to bring about an action potential, a wave of depolarisation will spread along the nerve fibre in both directions from the point of origin. This is how a nerve impulse travels along a nerve fibre.

10. The aim now is for you to draw a series of small diagrams to illustrate this chain of events. Each diagram will show a small piece of nerve axon. By the end of this section you'll have followed these changes – in effect the nerve impulse – from one end of the axon to the other.

    First copy out Figure 2.8A which shows a resting axon, with the outside of the membrane more positive than the inside. (I've omitted Na⁺ and K⁺ for the sake of clarity, but always be aware of how those ions move across the membrane at each stage of the action potential.)

11. Stimulation is applied to the axon membrane (whether by heat or pressure is irrelevent here) and immediately the membrane becomes much more permeable to sodium ions, which rush into the cell, as in Figure 2.8B. How will this affect the voltage across the membrane?

12. Local currents are set up between the depolarised and resting areas of the nerve fibre. The current flows in through the depolarised membrane and out

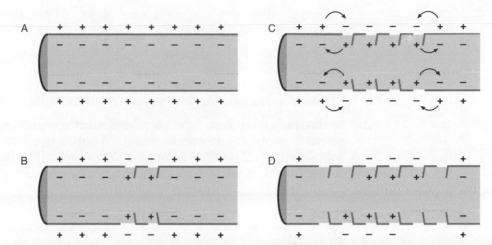

Fig. 2.8 (A) Resting state of nerve fibre. (B) Stimulus applied to membrane causes increase in permeability to Na+. (C) Propagation of action potential showing local currents. (D) Continuing propagation of action potential – add arrows to your diagram showing local currents, and changes in electrical charge.

through the adjacent resting membrane, as shown by the arrows in Figure 2.8C. The outward movement of the current through the membrane changes its permeability to Na+, which immediately rushes in. In this way the action potential is propagated in both directions along the nerve fibre from the source of stimulation.

13. Complete Figure 2.8D showing local currents and changes in electrical charge.

Pause here and look back at your copies of Figure 2.8A to D. Just remind yourself of the movement of sodium ions during the spread of the action potential.

14. Now find out how the subsequent repolarisation spreads along the fibre (in the wake of the spread of depolarisation) by using a new series of diagrams.

15. What does the term refractory period mean, and what is its significance for the conduction of a number of action potentials?

16. What is the difference in speed of conduction between myelinated and non-myelinated nerve fibres? Draw a diagram of a myelinated fibre showing the nodes of Ranvier. Explain saltatory conduction and its implication for the speed of conduction of a nerve impulse. See *Rutishauser 1994 Fig. 21.5*.

For a good readable summary of these events try *Guyton 1991, pp. 57–58*. Note, however, that he uses the unit milliequivalent for sodium and potassium concentrations rather than millimole.

 Take a well-deserved break here.

SECTION 2: THE SYNAPSE AND NEUROTRANSMITTERS
Time for completion: about 2 hours

You'll recall that both sensory and motor pathways consist of at least two neurones. This Section is concerned with events at the point where one neurone ends and the next one begins. The name given to this junction between two neurones is the synapse.

17. First read about the structure of the synapse. See: *Rutishauser 1994, Ch. 21*, the introduction and *Fig. 21.7; Guyton 1984, p. 130, and Figs 8.8 and 8.9; Marieb 1992, pp. 360–361, and Fig. 11.17.*

    Draw a diagram of the synapse, and label its various features.

18. Make notes on how chemical transmitters are manufactured, stored, and released from the nerve terminal. What is the function of the mitochondria found around the nerve terminals? What actually causes the release of chemical transmitter? What factors cause an increased release of transmitter?

 COMMENTARY ON QUESTIONS 17 AND 18

A list of some of the known chemical transmitters is provided in *Rutishauser 1994, Table 21.2*. In order to get the hang of nerve transmission across the synapse it's perhaps best to keep to just one example of a chemical transmitter for the moment. I would suggest concentrating on acetylcholine. Also, be clear about the distinction between presynaptic and postsynaptic neurones.

19. Once released, acetylcholine floods into the synaptic gap. Describe its fate from this point. See *Rutishauser 1994, p. 375*, 'Fate of the released transmitter'.

20. What happens at the postsynaptic neurone membrane on the arrival of acetylcholine? Is it always the case that an action potential along the postsynaptic neurone will result from this?

 COMMENTARY ON QUESTION 20

In your notes you'll need to distinguish clearly between inhibitory and excitatory potentials, that is, whether an action potential is caused or inhibited in the second neurone. This will depend on changes in voltage across the cell membrane of the postsynaptic neurone. See *Rutishauser 1994, pp. 375–376*. You'll need to define the terms temporal summation and spatial summation, perhaps making diagrams to illustrate them. See *Rutishauser 1994, Fig. 21.9*. Try also *Hubbard & Mechan 1987, p. 62*; and, for more detail, *Guyton 1984 pp. 131–133*.

21. Now read about the transmitter noradrenaline. Whereabouts in the nervous system is it found, how is it secreted and what is its fate following secretion? (Note that American texts such as Guyton and Marieb refer to noradrenaline as norepinephrine.)

FURTHER STUDY

Once you feel you're clear about the role of chemical transmitters, it's very useful to read around the subject in greater depth. Try *Marieb 1992*, with its excellent diagrams, pp. *360–365*; then pp. *365–370* for more information on different transmitters.

You might also like to read about the effect of different drugs on nerve synapses. See *Rutishauser 1994, p. 378*; and *Marieb 1992, p. 369*.

Can nerves re-grow if they are damaged? That depends on how great the damage is and where it is on the neurone. See *Rutishauser 1994, p. 381*.

Take a break here.

SECTION 3: THE CONTROL OF BALANCE AND FINE MOVEMENTS

Time for completion: 1 to 1½ hours. Sections 1, 2, and 3 should take you about 1 whole 'college day' to complete. But because this is such a complex subject, don't be surprised if you need longer.

The ability of the human body to perform delicate and intricate movements is astonishing. Watch a dancer, a rock climber or a tennis player, and consider how many groups of muscles are at work, maintaining balance and bringing about exactly the right movement at the right time. Here we consider the role of the cerebellum in helping to achieve actions such as these.

22. Look back at your notes on the functions of the cerebellum and add a diagram showing its position in relation to the cerebrum, brain stem, and spinal cord. See *Rutishauser 1994, p. 359*, then *p. 467 and Fig. 27.13*.

23. Make notes on cerebellar function. See *Rutishauser 1994, pp. 467–468 and 489*.

24. Draw a diagram showing nerve pathways from the cerebellum to muscle spindles. See: *Rutishauser 1994, Fig. 27.14; Mackenna & Callander 1990, p. 277*.

25. Continue your reading to discover the effects of damage to the cerebellum. What is meant by intention tremor, and how would you recognise it in a patient?

26. Now explore cerebellar functions in other texts. I'd recommend *Guyton 1984*: first the paragraph on *pp. 177–178*; then move on to *pp. 184–187*, 'Coordination of Motor Movements by the Cerebellum', with perhaps the following section on 'The Basic Neuronal Circuit of the Cerebellum'. When going into greater detail, don't lose sight of your basic descriptions of cerebellar functions – they are your 'mental foundations' without which further study is much more difficult.

    The description of cerebellar functions in *Marieb 1992, pp. 400 and 484*, is disappointingly brief, but see *p. 399* for a photograph of the cerebellum. For a more detailed description, turn to *Hinchliff & Montague 1988, pp. 127–130*.

27. Re-read your notes on the basal ganglia (Work Sheet, Question 15). Make sure you know where the basal ganglia are situated. Extend your notes by reading *Rutishauser 1994, p. 467*; then *Guyton 1984, pp. 177, and 180–182*.

Marieb 1992, pp. 390–391, refers to the basal ganglia as basal nuclei. *Figs 12.11 and 12.12* are useful.

 This is a good place for a break. This Guided Study so far may have taken you a full 'college day'.

SECTION 4: AUTONOMIC CONTROL
Time for completion: about 3 hours – a whole morning or afternoon

So far we've been mostly considering nervous system functions that are under voluntary control. This is sometimes referred to as the somatic nervous system. But not all nerves come under our conscious control. Imagine that you're watching a horror movie on television. As the climax approaches, you realise that the palms of your hands are sweaty and that your heart is 'thumping'. You feel the hair on the back of your neck stand on end, and your skin is covered in goose pimples.

The part of the nervous system that controls these unconscious responses – that keeps the body running smoothly as if on automatic pilot – is the autonomic nervous system. Its motor part has two main divisions, the sympathetic and the parasympathetic. We'll look at them in turn a little later.

28. First, read the brief introduction to the autonomic nervous system in *Rutishauser 1994, p. 47*. It makes a very useful distinction between the autonomic and somatic systems, which it is good to bear in mind. Note the importance of the autonomic nervous system to the maintenance of homeostasis.

29. Now continue your reading, and make notes on the sensory and integrative parts of the autonomic nervous system (*Rutishauser 1994, pp. 47–49*).

 COMMENTARY ON QUESTION 29

Make sure you understand the terms ganglia and plexus – see *Fig. 3.14* in *Rutishauser 1994*. It's interesting that some texts seem to imply that the autonomic system is a motor system only. True, we'll be concentrating on the motor divisions of the autonomic nervous system, but you'll appreciate that there has to be an input of information from sensory nerves, which is examined and evaluated before commands are sent out.

Sympathetic

30. Look first at *Marieb 1992, pp. 459–460*, where brief descriptions of the roles of both sympathetic and parasympathetic divisions are given. Which of the divisions do you think is most at work in someone watching a horror movie? Note how Marieb underlines that, though the divisions are often described separately for the sake of clarity, they more often than not work together. We'll see examples of this later.

31. Draw a diagram showing the sympathetic nerve outflow from the spinal cord to . . . which important structure? See *Rutishauser 1994, Fig. 3.16* (right hand side of the figure, but note that the nerve outflows occur on both sides

of the spinal cord). Then show the various ganglia supplied by the nerves, and finally the organs that receive innervation from the sympathetic system. You could also have a look at: *Marieb 1992, Fig. 14.2; Guyton 1984, Fig. 12.2; Mackenna & Callander 1990, p 287.*

Ensure that you are clear about which sections of the spinal cord the sympathetic fibres leave.

32. Label the nerve fibres appropriately in your diagram: pre-ganglionic fibres and post-ganglionic fibres. (See *Guyton 1984, p. 193.*)

33. What is the connection between the sympathetic nervous system and the adrenal glands? Show this on your diagram (or a fresh one) and write brief notes.

34. What is the neurotransmitter at the synapse between the pre- and post-ganglionic fibres?

35. What is the effect of sympathetic stimulation on:

    - the airways
    - the heart muscle
    - the pupils
    - skeletal muscle blood vessels
    - blood glucose levels?

(See *Guyton 1984, Table 12.1, p. 196.*)

Is there anything that functionally links these changes? In other words, what is the point of them?

 COMMENTARY ON QUESTION 35

You may have come across the phrase: 'fright, fight or flight' to describe the point of sympathetic activity (see *Hubbard & Mechan 1987, p. 229*). In other words, sympathetic stimulation helps us stand and fight, or turn and run away. In either case, our heart and respiratory rates increase, and the airways dilate so that we can get more air into our lungs.

So why does the hair on the back of the neck stand on end? (Think about how a dog's hair stands when it is suspicious or angry. Does that serve any purpose in the animal world?)

Think also about the effects of sympathetic activity on the intestines, both gut movement and sphincter control, and on the bladder. What are these effects, and is there any connection with 'fright, fight or flight'? (Don't be confused by the fact that extreme fright can cause sudden bowel movement – this is a specialised activity of the parasympathetic system.)

---

36. *Guyton 1984, pp. 195–198,* provides a good description of the effects of both sympathetic and parasympathetic systems on various organs. This passage will stand revisiting after the next Section of this Guided Study.

37. Try to summarise your knowledge of the sympathetic nervous system by showing, in diagram form, the appropriate nerve supply from the spinal cord to just three of the various affected organs. I'd suggest the heart, lungs, and blood vessels. (I'd leave the adrenal glands until the appropriate Guided Study on the endocrine system.)

Parasympathetic

Here we follow very much the same pattern of investigation as with the sympathetic system. Unfortunately, there's no convenient phrase like 'fright, fight or flight' to describe parasympathetic effects; but from what you've read so far, how might you sum up the point of parasympathetic activity?

38. Draw a diagram showing the parasympathetic nerves supplying various body organs. Show where these nerves leave either the brain or the spinal cord, and note how this outflow differs from that of the sympathetic system. See: *Rutishauser 1994, Fig. 3.16; Marieb 1992, Fig. 14.2* (right hand side); *Mackenna & Callander 1990, p. 286; Guyton 1984, Fig. 12.3.*

39. Again, distinguish on your diagram between pre- and post-ganglionic nerves. How do parasympathetic pre-ganglionic fibres differ in length from their sympathetic counterparts? How do the two sets of post-ganglionic fibres differ?

40. Make notes and diagrams on the effect of parasympathetic stimulation to the intestines, salivery glands, and urinary bladder.

41. Construct a diagram showing both parasympathetic and sympathetic supplies to the heart. Demonstrate how these nerve supplies work together to maintain a healthy heart beat under differing circumstances such as rest, mild exercise, and strenuous exercise.

42. Re-read the overview of the effects of both nerve supplies to various body organs in *Guyton 1984, pp. 195–198.* You might also try *Hubbard & Mechan 1987, pp. 229–231.*

43. Read *Guyton 1984, p. 194,* on the different transmitters. *Rutishauser 1994, Table 3.7,* shows how our knowledge of such transmitters is increasing, but it may be sensible to concentrate first on acetylcholine and noradrenaline. What do cholinergic and adrenergic mean? How does the sympathetic stimulation of sweat glands differ from that of other organs? (Think about the neurotransmitter involved.)

44. Finally, choose one or two organs other than the heart and draw diagrams showing how the sympathetic and parasympathetic supplies work together. Write notes to explain your diagram(s).

 FURTHER STUDY

We have some degree of conscious control over our respiration rate and, by exercising hard, we can make our heart rate increase. But can we consciously reduce it, or lower our blood pressure? Well, just by thinking, we can't, any more than we can increase or decrease the flow of pancreatic juices. But biofeedback techniques are now being used – either as well as or instead of drugs – to help people reduce their blood pressure. Relaxation exercises include concentration on breathing, so that the respiration rate slows down. It is claimed that, as the breathing rate slows, so does the heart rate, and the blood pressure falls. The advantages to the patient are that he can avoid drugs and their side effects, while contributing positively to his treatment and well-being.

Find out about biofeedback techniques, paying particular attention to their effect on physiology. Start with *Marieb 1992, p. 472.*

SECTION 5: BLOOD SUPPLY TO THE BRAIN
Time for completion: 1½ to 2 hours

45. Draw a diagram showing the arteries that leave the aorta to supply the head and neck. Show also the venous return (into which major blood vessel?) from the head to the heart.

46. Find out what is the circle of Willis, and draw a diagram illustrating it.

The supply of blood to the brain is usually maintained despite a person's widely changing levels of activity, from lying down to rest, standing for long periods, or running a marathon. The reason for this stable supply is that the brain – of all body organs – relies on a constant supply of oxygen in order to function.

 ACTIVITY

Discuss with colleagues, and perhaps with your tutor, what happens when someone faints. What can cause a person to pass out, and what happens to his or her blood pressure? What should be the immediate first-aid treatment for the victim of a faint, and why?

You might like to practice taking each other's blood pressure, with the 'patient' first standing, then sitting, then lying flat. How much does the blood pressure vary?

47. Write notes on the regulation of blood supply to the brain. See, for example: *Rutishauser 1994, pp. 362–364; Guyton 1984, pp. 294–295; Marieb 1992, pp. 649–650.*

 COMMENTARY ON QUESTION 47

You will need to explain the term autoregulation. You should refer to the local effects of changing levels of carbon dioxide in the arterial blood supplying the brain, and how this is monitored. (Are changes in oxygen content as important as changes in carbon dioxide?) How are fluctuations in blood pressure monitored within the body?

Arterioles throughout the body have a sympathetic nerve supply, which influences their diameter and so affects the flow of blood through them. How does the sympathetic supply to cerebral arterioles differ from the supply to other arterioles in the body?

The capillaries within the brain are much less permeable than capillaries elsewhere in the human body. They allow certain substances through to the brain, but prevent the entry of others. This is called the blood–brain barrier.

48. Read and make notes about the blood–brain barrier. State which substances can cross it, and how, and which are prevented. See: *Rutishauser 1994, p. 363; Hubbard & Mechan 1987, p. 250; Jennett 1989, p. 54; Marieb 1992, p. 405.*

See *Rutishauser 1994, Table 20.3,* for a list of substances that cross the barrier.

FURTHER STUDY

In meningitis (inflammation of the meninges, or protective membranes covering the brain) treatment with antibiotics, such as penicillin, is essential. Try to find out how easily penicillin crosses the blood–brain barrier. What are the implications for achieving the correct dosage?

49. Finally, if you have time, read about the production of cerebrospinal fluid (CSF) and its flow through the ventricles of the brain. See: *Wilson 1990, pp. 247–248; Mackenna & Callander 1990, p. 127; Rutishauser 1994, pp. 364–365; Guyton 1984, pp. 352–353; Marieb 1992, pp. 404–405.*

Make a list of the composition of CSF, and see how it compares with the composition of blood.

■ Pain (suggested reading)

You have already studied nerve endings and pathways, and how nerve impulses are transmitted to the brain. You have learned how the body can make swift reflex movements to withdraw limbs from a source of potential pain and tissue damage.

Use the following references to read about, and make notes and diagrams on:

- tissue receptors and causes of pain
- types of pain – fast and slow – and nerve fibres
- the gate control system
- endogenous or natural analgesia
- referred pain
- common types of pain such as headache and toothache
- forms of pain relief including drugs.

Suggested reading

Roper, Logan, & Tierney 1990, pp. 121–125. This will provide a good introduction for further reading.

Then go on to: *Rutishauser 1994, Ch. 32* (but be prepared to follow up references to other relevent chapters); *Boore, Champion, & Ferguson 1987,* 'Appendix: The Physiology of Pain' (some excellent diagrams but this reference goes into much greater depth than Rutishauser).

You may also like to dip into one of the 'classics' on pain: *Melzack & Wall 1988.* Look at Part One; 'The Puzzle of Pain'. This will certainly whet your appetite for studying the physiological mechanisms of pain. Why, for example, can someone feel pain where there is no physical injury? Why can quite dramatic injury occur without a person feeling any pain at all? What is the placebo effect? Does it mean that a person's pain isn't genuine?

It's important to appreciate that pain is not only a physiological process. The experience of pain involves the emotions, one's upbringing and social, racial and educational background, and any past experience of pain.

Think for a moment of your last visit to the dentist. If that experience of dental treatment (rather than just a checkup) was pain free, you are likelier to be a little more relaxed on your next visit, than if the earlier experience was relatively painful. One's state of mind – stressed and apprehensive, or cheerful and relaxed – greatly affects the experience of pain, and the way in which one reacts to it.

If you have time, it is helpful to read of people's experiences of, and reactions to, pain of various types. My first disability article (Goodall 1988) was written because I wanted to explain the effects of chronic pain on my lifestyle, and how I responded.

REFERENCES
Boore J, Champion R, Ferguson M 1987 Nursing the physically ill adult. Churchill Livingstone, Edinburgh
Goodall C 1988 Living with pain. Nursing Times 84 (32): 62–63
Melzack R, Wall P 1988 The challenge of pain, revised edn. Penguin Books, Harmondsworth
Roper N, Logan W, Tierney A 1990 The elements of nursing, 3rd edn. Churchill Livingstone, Edinburgh

■ The nervous system (check quiz)

1. Draw a diagram showing the position of sensory receptors in the skin. Describe how a nerve impulse travels from one of these receptors to the spinal cord, and thence to the brain.

2. What is a motor unit?

3. Distinguish between the terms depolarisation and repolarisation. What is meant by an all or nothing response?

4. Describe the events at a synapse following the arrival there of a nerve impulse. How is acetylcholine produced, and what is its fate?

5. Sue's resting pulse rate is about 58 beats per minute. During an ascent of Glyder Fawr in Snowdonia, she notices that her pulse rises to 72, then 96, and finally to 108 beats per minute. In this situation, what are the 'accelerating' and 'braking' effects of the autonomic nervous system on her heart? What other parts of her body will be affected by the autonomic nervous system?

6. Discuss the following situation with your colleagues.

    Simon was involved in a car crash 6 months ago, when he suffered a fractured spine. His spinal cord was completely severed at the level of the 1st and 2nd lumbar vertebrae. As a consequence he is unable to move his legs, though he has complete arm movement.

    Will Simon be able to feel pressure, touch or temperature in his legs and feet? If your answer to this is 'no', what implications will there be for his future care, either in hospital or his home?

    Will there be any reflex movements in his feet and legs?

7. Attempt this question if you tried some of the suggested reading on pain.

    Explain the main differences between chronic and acute pain. How do you think a nurse's knowledge of the physiology of pain will help her or him in the care of a patient in pain?

8. In physiological terms, what is memory?

Introduction to endocrines — thyroid and parathyroid glands (work sheet)

Time for completion   About 2 to 3 hours

Overall aim   To achieve a basic understanding of the role of endocrines in the body – their production, actions, and regulation of production – by studying the thyroid and parathyroid glands.

Introduction   The endocrine system is one of the communication or regulatory systems of the body. Whereas the nervous system is capable of responding quickly to changes, the endocrine system tends to work by maintaining, inhibiting or exciting the state of the body over a period of time.

Hormones, the chemicals secreted by endocrine glands, are sometimes described as chemical messengers. They are secreted into the surrounding tissue fluid or into the blood, and are conveyed to the target organ where they have their effect.

The chemical structure of hormones will be covered in the following Guided Study.

Background questions
1. Distinguish between an endocrine gland and an exocrine gland. What is the main structural difference between the two types of glands? (Use a diagram if you think it will clarify your notes.) Give two or three examples of each.

2. Draw a labelled diagram showing the position of the principal endocrine glands. See: *Wilson 1990, Fig. 14.1, p. 311; Rutishauser 1994, Fig. 3.8, p. 41; Marieb 1992, Fig. 17.1, p. 541.*

   Read the section in *Rutishauser 1994, p. 41,* describing the location of endocrine glands, and read the introductory paragraphs to *Ch. 10.*

3. The opening pages of *Guyton 1991, Ch. 74,* provide a good introduction to the endocrine system and its various glands. Try the sections: 'Nature of a hormone', *p. 810;* 'An overview of the important endocrine glands', *pp. 810–811;* and 'Storage and secretion of hormones', *pp. 812–813.*

Specific questions   **THE THYROID GLAND**
Time for completion: 1 to 1½ hours

1. Draw a detailed diagram showing the position of the thyroid gland and important structures nearby. See: *Wilson 1990, Fig. 14.6, p. 315; Rutishauser 1994, Fig. 10.11A, p. 201.*

2. Describe the thyroid tissue, perhaps using a new diagram showing thyroid cells and follicles.

3. What is colloid, and how is it produced?

4. Two of the hormones produced by the thyroid gland are thyroxine and tri-iodothyronine (otherwise referred to as T_4 and T_3). What dietary ingredient is required for the synthesis of these hormones? In which foodstuffs is this substance found? Read about the synthesis of the thyroid hormones in *Rutishauser 1994, pp. 200–201.*

5. Summarise the actions of T_3 and T_4 (Commentary, p. 64).

COMMENTARY ON QUESTION 5

However simply you choose to answer Question 5, be certain that you understand the terminology you use. For example, thyroxine is said to stimulate the metabolic rate, but what exactly does this mean?

You've read about the synthesis of certain substances (including T_3 and T_4). In fact, you should be 'synthesising' your answers to these questions – assembling the information from various sources and then writing it down in a short yet fully comprehensible form. In this way you stand a better chance of learning than by simply copying whole sentences or paragraphs from your textbooks.

As well as looking at Wilson 1990 and Rutishauser 1994, you could try the very full account of the functions of thyroid hormones in *Hubbard & Mechan 1987, pp. 308–309.*

Then try *Guyton 1991, pp. 835–836*, where the effect of thyroid hormones on the body systems is described.

REGULATION OF THYROID HORMONE PRODUCTION
Time for completion: about half an hour

6. How is the production of T_3 and T_4 regulated? In other words, what increases and decreases the secretion of these thyroid hormones?

 COMMENTARY ON QUESTION 6

First look at the role of the anterior pituitary gland, and its production of thyroid stimulating hormone (TSH) or thyrotrophin (thyrotropin in American texts). Then read about how the hypothalamus controls the secretion of TSH.

You don't at this stage need to study the structure of the pituitary, or any of its other roles, since they are covered in the following Guided Study.

You may find that a flow diagram helps you understand the relationship between hypothalamus, anterior pituitary, and thyroid gland, and their various secretions.

See: *Wilson 1990, pp. 315–316; Rutishauser 1994, p. 202; Hubbard & Mechan 1987, pp. 309–310; Marieb 1992, p. 551; Guyton 1991, pp. 836–838.*

7. Look again at your notes on the control of thyroid hormone secretion, and check any diagram you've drawn. What you have described is an example of a negative feedback system. Why negative? You'll find other examples of feedback systems in the next Guided Study. Read *Rutishauser 1994, p. 34*, for a review of negative feedback and its role in homeostasis. Ensure that you understand that important term.

 Take a short break here.

THE PARATHYROID GLANDS
Time for completion: about 1 hour

8. Identify the position of the parathyroid glands and note their proximity to

other structures (including the thyroid gland). See, for example, *Wilson 1990, Fig. 14.9, p. 316.*

9. What size are the parathyroid glands? How easily do you think they may be identified during surgery on the thyroid gland? Textbooks frequently state that there are four glands, but unfortunately for the surgeon the number can vary from two to ten. See *Rutishauser 1994, p. 204, and Fig. 10.11B.*

The hormone secreted by the parathyroids is parathormone (also known as parathyroid hormone) and its secretion varies according to the level of calcium in the blood.

10. What is the normal range of blood calcium?

11. How does a raised level of calcium in the blood (hypercalcaemia) or a lowered level of calcium (hypocalcaemia) affect parathormone production?

12. The secretion of parathormone increases the level of calcium in the blood, but where does this additional calcium come from? What might be the effect of long-term oversecretion of parathormone? (Think about a source of calcium within the body.)

 FURTHER STUDY

During surgery to the thyroid gland, the parathyroids may be inadvertently removed or damaged. This will lead to a reduction in blood calcium levels. Find out what clinical effects this might have, and what observations should be made by the nurse on a patient following thyroid surgery.

A third hormone produced by the thyroid gland also affects blood calcium levels. This hormone is called calcitonin. However, it has the opposite effect on calcium levels to parathormone.

13. Draw a simple diagram showing the relationship between parathormone, calcitonin, and blood calcium levels. Compare your diagram for accuracy and clarity with those of your colleagues.

(Have a look at *Hubbard & Mechan 1987, Fig. 9.11, p. 312,* but not until you've worked out your own diagram first.)

 FURTHER STUDY

It sometimes helps to clarify the functions of a hormone if we study what happens when that hormone is secreted in too great or too small a quantity. Discuss with colleagues, now that you know the actions of thyroxine, how a person might feel or behave with excessive or reduced T_4 production. Then find a nursing text which describes hyperthyroidism and hypothyroidism, and have a look at the photographs in *Guyton 1991, Figs 76.7 and 76.8, pp. 839 and 840.*

■ Further exploration of the endocrine system (guided study)

Time for completion   Between 8 and 10 hours, or 1½ 'college days'

Overall aim   To continue the study of the endocrine system by examining the adrenal glands and the pancreas; and to explore the role of the hypothalamus and pituitary gland in the control of hormone production.

Introduction   In the preceding Work Sheet we examined the thyroid and parathyroid glands, as examples of how hormones are produced and the effects they have on distant tissues and organs. In the course of our studies we discovered that certain hormone secretion was controlled by the pituitary gland and the hypothalamus.

To begin one's work on the endocrine system by studying the pituitary gland (because of its central importance to the endocrine system) seems to me almost to invite difficulties. For this reason I preferred to examine the production and actions of two less central hormones, so that one begins to understand, through them, the principles underlying the system as a whole. It is the purpose of this Guided Study to add more detail to this basic understanding.

Background questions

1. Read *Jennett 1989, pp. 36–40*, for an interesting historical overview of the study of hormones. (Ignore for the moment the Table of principal hormones on pp. 38 and 39.)

2. Check your earlier diagram (from the preceding Work Sheet) showing the position of endocrine glands, noting especially the positions of:

   - the pituitary gland
   - the adrenal glands
   - the pancreas.

3. Read *Rutishauser 1994, Ch. 3*, about the position of endocrine tissue and the chemical composition of hormones. Note that endocrine tissue can be grouped into discrete glands with their own blood supply (e.g. the thyroid gland) or situated within other organs such as the pancreas, lungs, and intestines.

   There are two main chemical groups of hormones, peptides and steroids. (You have probably heard of drugs called steroids.) See *Rutishauser 1994, Fig. 3.9*, for lists of hormones belonging to each group. Note that thyroxine is neither a peptide nor a steroid. Read Rutishauser's description of how steroids and peptide hormones act on cells (*pp. 26 and 42*). See also: *Jennett 1989, p. 40; Hubbard & Mechan 1987, pp. 297–300; Marieb 1992, pp. 541–545* (more detailed but with excellent diagrams).

Specific questions

SECTION 1: THE ROLE OF THE HYPOTHALAMUS AND PITUITARY GLAND
Time for completion: between 2 and 3 hours

1. First read your notes on the roles of thyroid stimulating hormone (TSH) and the hypothalamus in the production of thyroxine. Check your notes on how the hypothalamus acts on the anterior pituitary, which in turn acts on the thyroid gland, and make sure you understand the term negative feedback system.

2. Show in diagram form the position of the pituitary gland, both anterior and posterior lobes, in relation to the hypothalamus. Read about the different development of the anterior and posterior lobes, and the type of cells that make up each lobe. Note the blood supply to the anterior lobe of the gland. See: *Wilson 1990, Figs 14.2 and 14.3; Hubbard & Mechan 1987, Fig. 9.3; Rutishauser 1994, Fig. 10.5; Marieb 1992, Fig. 17.5.*

3. Find out the alternative names for the anterior and posterior lobes of the pituitary gland.

 ACTIVITY

How big is the pituitary gland? Despite its many important functions, it is remarkably small. Using a skull from your college, try to locate the small hollow (the hypophyseal fossa) where the pituitary is located. (*Wilson 1990, Ch. 16*, may help you, and *Marieb 1992, p. 186.*)

We'll concentrate first on the anterior pituitary, and the network of blood vessels connecting it with the hypothalamus.

4. What hormones secreted by the anterior pituitary influence hormone production from:

   - the thyroid gland
   - the adrenal cortex?

(You may like to note the names of the other principal anterior pituitary hormones, including the sex hormones, but we'll not be considering them here.)

5. How does the hypothalamus control the secretion of the anterior pituitary hormones you've just listed?

 COMMENTARY ON QUESTION 5

You will need to discover the names of the releasing hormones from the hypothalamus, and how they are produced. See *Rutishauser 1994, Ch. 10*; also *Hubbard & Mechan 1987, pp. 301–302.*

Try to express the relationship between hypothalamus, anterior pituitary, and target tissue, in diagram form, then check your diagram with *Hubbard & Mechan 1987, Fig. 9.4.* You may find it clearer to draw separate diagrams for each anterior pituitary hormone.

Do your diagrams show any form of feedback mechanism?

---

6. *Rutishauser 1994, p. 190,* describes the hypothalamus as the 'interface between the nervous system and the endocrine system'. Do you understand this? Read what Rutishauser says about sleep and the production of hypophysiotrophic hormones, and also the effects of stress.

FURTHER STUDY

The physiological effects of stress are widespread, affecting just about every body system. Why not start a special section in your notes headed 'Stress' and, as you study each system of the body, add relevant information to it. By the end of your course you should have amassed much useful material which will be valuable if you are interested in sports physiology. (See *Guyton 1984, Ch. 39.*)

7. Read about the production of growth hormone (GH) from the anterior pituitary, and its effects on proteins, fats and carbohydrates. What is the role of somatomedin? Does growth hormone directly affect the development of bone and cartilage? See: *Rutishauser 1994, pp. 194–195; Guyton 1984, pp. 564–567; Marieb 1992, p. 550* (photographs showing the effects over time of oversecretion of growth hormone; note the changes to the hands, nose and jaw).

 FURTHER STUDY

You might like to find out the effects of undersecretion and oversecretion of growth hormone during childhood.

While one can study physiology without reference to disease, examining the effects of over- and undersecretion of various hormones helps to clarify the endocrine system.

8. Check that you are clear about the way in which the posterior pituitary gland is connected with the hypothalamus. Read about the secretion of antidiuretic hormone (ADH) from the posterior pituitary. See: *Wilson 1990, p. 314; Rutishauser 1994, pp. 193 and 242; Hubbard & Mechan 1987, pp. 305-307; Guyton 1984, pp. 569–570.*

Note how ADH is formed and secreted, and what physiological changes influence its secretion.

9. Draw a diagram showing the links between the osmotic pressure of blood and ADH secretion (then compare your diagram with *Wilson 1990, Fig. 10.10*). What will happen to ADH production if you take no fluid for several hours? And what if you drink a lot within a short time? You may like to express these events in diagrams.

10. What is the alternative name for ADH, and what physiological action does this name suggest?

 Take a break here. But read through your notes first to make sure you fully understand them.

SECTION 2: THE ADRENAL GLANDS, CORTEX AND MEDULLA
Time for completion: 3 or 4 hours

The adrenal glands are sometimes called the suprarenal glands because of their position in relation to the kidneys. The adrenal medulla and cortex may be regarded as distinct glands with separate functions because they develop from different types of embryonic tissue. However, hormones from both parts of the adrenals help the body to respond to stress.

11. Check your earlier notes on the position of the adrenal glands in relation to the kidneys. Draw a cross-section of an adrenal gland, showing the outer layer (cortex) and the inner layer (medulla).

The adrenal medulla
Time for completion: about 1 hour

The adrenal medulla secretes two hormones, adrenaline and noradrenaline (called epinephrine and norepinephrine in American texts). Both these hormones are called catecholamines.

12. Make notes on how the sympathetic part of the autonomic nervous system is anatomically and physiologically linked with the adrenal medulla. A diagram would be useful here.

13. From which substance (amino acid) are the catecholamines formed?

14. What causes the secretion of adrenaline and noradrenaline, and which is secreted in the greater quantity?

15. Describe the effect the two catecholamines have on:

   - heart muscle
   - coronary arteries
   - respiratory rate
   - skeletal muscle.

Later in your nursing career you'll perhaps come across certain drugs described as sympathomimetic. Find out what this term means – it's a very useful one because it helps us understand how closely linked the adrenal catecholamines are with the sympathetic nervous system. (See *Guyton 1991, pp. 270–271*.)

16. Find out about the significance of the alpha and beta adrenoreceptors found in different organs and, in particular, in the coronary arteries (*Rutishauser 1994, Tables 10.4 and 10.5*).

 COMMENTARY ON QUESTION 16

Consider what effect sympathetic stimulation has on the size of blood vessels. Those in the skin will constrict, but the coronary arteries dilate, both with sympathetic stimulation, and with the release of adrenaline. Why do they behave differently? Why do the coronary arteries dilate when the release of noradrenaline (rather than adrenaline) would cause them to constrict? (Think about the different concentrations of both these catecholamines released from the adrenal medulla.) See *Rutishauser 1994, p. 198*.

---

17. Sympathetic stimulation provides a swift response to stress. Find out how catecholamine release compares in time of effect. Which acts more quickly, and which lasts longer?

18. What is the effect of catecholamine release on blood glucose levels? See *Hubbard & Mechan 1987, p. 322*. Note the meaning of the terms gluconeogenesis and glycogenolysis.

 Take a short break here.

The adrenal cortex
Time for completion: about 2 hours

19. Name the three layers or zones of the adrenal cortex, noting which zone is the largest.

In this Guided Study we'll concentrate on two important hormones produced from the adrenal cortex: cortisol and aldosterone.

20. From which parts of the cortex are these hormones produced? Read how these two steroid hormones are synthesised, and how they are carried in the blood to their target tissues. (See *Rutishauser 1994, p. 198*.)

First we'll consider the mineralocorticoid effect of, especially, aldosterone; that is, the movement of sodium and potassium across cell membranes, thus controlling their retention or excretion from the body.

21. How does aldosterone act on the kidney tubules, and what effect does it have on sodium levels in the blood? Find out also what effect aldosterone has on blood levels of potassium. See: *Rutishauser 1994, p. 245; Hubbard & Mechan 1987, pp. 319–320; Guyton 1984, pp. 580–581; Marieb 1992, p. 560*.

22. What comparative effect does cortisol have on sodium and potassium balance? Why does it have such an effect?

23. Describe how the secretion of aldosterone is controlled. Make use of a diagram to show the various factors at work here.

 COMMENTARY ON QUESTION 23

You will need to include factors such as the release of ACTH from the anterior pituitary, and, especially, angiotensin II from . . . where?

What effect do serum levels of potassium and sodium have on aldosterone secretion?

Try a number of texts for this quite difficult topic, in increasing order of complexity: *Wilson 1990, p. 317, and Fig. 14.11B; Rutishauser 1994, pp. 200 and 245, and Fig. 13.7; Hubbard & Mechan 1987, p. 320, and Fig. 9.17; Guyton 1984, pp. 581–582; Marieb 1992, pp. 560–562, and Fig. 17.13*.

Now we turn to cortisol and its powerful glucocorticoid action. You've already discovered that cortisol has a slight effect on sodium and potassium levels in the blood; but its principal action is on blood glucose.

24. Cortisol secretion causes blood glucose levels to rise. Find out how this is brought about. You'll discover that cortisol is also concerned with the metabolism of proteins and fats, so make notes on the effect of cortisone on glucose, protein and fat metabolism. (See *Guyton 1984, pp. 582–583*.)

25. How is the production of cortisol controlled? Is cortisol secretion constant, or is there any daily alteration? (See *Rutishauser 1994, Fig. 18.1*.) What sort of events might bring about a rise in cortisol secretion? What is the link between cortisol production and stress? See *Rutishauser 1994, p. 339*.

26. What effect does cortisol have on the immune system especially when secreted in large amounts? See *Hubbard & Mechan 1987, pp. 318–319*.

 FURTHER STUDY

Patients receiving steroid therapy can suffer side effects because they are less able to withstand infection. Once you have clear notes on the physiological action of cortisol, you may like to read about the problems and nursing responsibilities of steroid medication. See, for example, Hopkins 1992. Look under drugs such as prednisolone or hydrocortisone.

---

27. Try to draw a diagram that summarises the response of the adrenal glands – medulla and cortex – to stress. See: *Guyton 1984, pp. 583–584; Marieb 1992, Fig. 17.14, p. 564; Rutishauser 1994, Fig. 19.4, p. 339; Hinchliff & Montague 1988, Fig. 2.5.29, p. 196*.

    Can you suggest how these stress responses might help to bring about a normal physiological state (homeostasis)? Can this be shown on your diagram?

    Think, for example, about loss of blood. This will lead to a fall in blood volume, a fall in blood pressure, a rise in the pulse rate . . . Now, how will hormone production from the adrenal medulla and cortex help to bring about a return to a near-normal blood volume and pressure? (This may be useful as a discussion topic with colleagues, using a white board and markers so that each of you can try out your ideas as you 'think aloud' – always a useful exercise.)

    Read *Rutishauser 1994, pp. 270–272*, for a description of endocrine and other responses to a period of starvation.

28. Note also that male sex hormones (androgens) are produced from the adrenal cortex, in both males and females. Briefly read about these and note the comparatively slight effect they have compared with testosterone formed in the testes. See *Hubbard & Mechan 1987, pp. 321–322*.

 Take a substantial break here.

SECTION 3: HORMONES FROM THE PANCREAS
Time for completion: about 3 hours

As well as secreting digestive enzymes (see Unit 4) the pancreas has endocrine cells which secrete a number of hormones. Here we'll be considering principally two of these – insulin and glucagon. You've probably heard of insulin in relation to the condition known as diabetes mellitus.

29. Make sure that you know the difference between endocrine and exocrine cells. What are the islets of Langerhans? List the different types of endocrine cells within the pancreas, and the hormones they produce.

    Both insulin and glucagon are concerned with metabolism. At first it's easy to think of these hormones as involved only with the metabolism of carbohydrates, but in fact proteins and fats are affected too.

30. Insulin helps glucose to cross the cell membrane in most tissues. (Glucose is the end-product of the metabolism of what foodstuffs?) Note the terms insulin-dependent tissues and insulin-independent tissues. Which of the tissues in the body are insulin independent? (See *Rutishauser 1994, pp. 205–206.*)

31. What is the normal range of blood glucose? (Don't forget to note the units used.) The blood glucose level is actually a measure of extracellular glucose – the amount of glucose within the plasma, and outside the cells. Consequently, as glucose moves inside cells, the blood glucose level will fall.

32. Bearing this in mind, what effect does insulin have on blood glucose? Does the level rise or fall?

33. What stimulates insulin secretion? Is it higher before or after a meal?

For Questions 30 to 33 see: *Wilson 1990, p. 318; Rutishauser 1994, pp. 206, and 267; Guyton 1984, pp. 586 and 588–589; Marieb 1992, p. 566, and Figs 17.15 and 17.16.*

COMMENTARY ON QUESTIONS 30 TO 33

Think of this in stages. As you get out of bed first thing in the morning, will your blood sugar be high or low? Now you have breakfast, perhaps with cereals, milk and sugar, toast and marmalade. As these foodstuffs are digested, what will happen to your blood sugar level – will it rise or fall? As the blood sugar level changes, so will the concentration of glucose in the tissue fluid surrounding the pancreatic cells that produce insulin. It is this changing glucose concentration that either stimulates or inhibits the secretion of insulin.

---

34. Glucagon has the opposite effect on blood glucose to insulin. Read about the production of glucagon, and what causes its secretion.

35. Draw a diagram showing the effects of insulin and glucagon on blood glucose. Try also to show factors that stimulate the secretion of these hormones, and any feedback mechanisms you think are appropriate.

36. Glucose in the diet is useful for providing energy quickly. (Why do you think it might be more useful than, say, ordinary sugar?) A balanced diet should contain sufficient energy-providing foods for our daily needs, but what happens when we ingest too much carbohydrate? Read about the effects of insulin on excess glucose that result in its conversion to:

    - glycogen in the liver and muscles
    - fat which is then stored in various parts of the body.

(See *Guyton 1984, pp. 586–587.*)

37. Make notes on the effects of insulin on the metabolism of both fats and proteins. See *Guyton 1984, pp. 587–588;* then turn to *Rutishauser 1994, Ch. 14,* for a review of the hormonal control of metabolism. Note what happens to insulin and glucagon secretion when a meal is taken, and when fasting occurs, including extended periods of fasting. Check that you understand the terms glycogenolysis and gluconeogenesis.

 FURTHER STUDY

An obvious topic for further study is the condition known as diabetes mellitus, where very little or no effective insulin is produced. This is a highly complex subject, so you may like to just 'dip' into a textbook of medical nursing which will provide you with a basic overview.

38. From the work you've now completed on endocrines, do you feel able to explain the term homeostasis, and give examples? Write a summary of your understanding of homeostasis.

39. Finally, check your notes on stress that I suggested you make. What information do you have about how the endocrine system responds to stress, including exercise?

REFERENCE
Hopkins S J 1992 Drugs and pharmacology for nurses, 11th edn. Churchill Livingstone, Edinburgh

■ The endocrine system (check quiz)

1. Explain the term homeostasis. Can you think of an example of homeostasis to which endocrines contribute?

2. Choosing two hormones you have studied, draw a diagram demonstrating how the hypothalamus and the pituitary gland control the secretion of those hormones. Add to your diagram any feedback mechanism that is appropriate.

3. A colleague tells you he is confused between exocrine and endocrine glands. Outline the main differences between them, using the pancreas as your example.

4. Endocrines help maintain the body's fluid balance. Using diagrams where appropriate, describe how the endocrine system would respond to these two situations:

   a. A young man drinks five pints of beer during an evening's celebrations.
   b. A patient is kept 'nil by mouth' (i.e. without food or drink) for 6 hours preceding her operation.

   (First decide what endocrines might be involved in each case; then work out how they would react to these situations.)

5. Endocrines also play an important part in helping the body respond to stressful situations. For example, Jill is revising hard for her final exams in nursing, due in 2 months' time. How might her endocrine system respond to this lengthy period of stress? (You might also like to consider how her nervous system would play its part too.)

   In particular, think about the contribution of the adrenal glands, both cortex and medulla. When considering the adrenal medulla, work out the input from the sympathetic nervous system.

6. Finally, here is an example of a relatively short period of stress. Alan is a keen wheelchair athlete and is hoping to beat his personal best time of 1 hour and 59 minutes in the forthcoming London Marathon. How will his endocrine system help his body cope with this very stressful, if relatively short, period of exercise?

   (Think first about the effect of the exercise on his blood sugar, and the subsequent reaction of his pancreas, adrenal medulla, and adrenal cortex. You might also like to consider how the nervous system would help Alan to function efficiently during the race. What will happen to his heart rate, his blood pressure, his respiratory rate, his muscle blood flow? Check through your notes on stress that I suggested you make – do they help you answer some of these questions?)

■ The special senses — hearing and sight (suggested reading)

HEARING

Use the following references (arranged in order of increasing complexity) to make notes and diagrams about:

- the nature of sound, its pitch and intensity
- the role of the external ear in 'collecting' sound
- the role of the middle ear in transmitting vibrations
- the function of the Eustachian tube (or pharyngotympanic tube)
- the role of the inner ear in converting vibrations to nerve impulses
- sending nerve impulses to the brain
- the perception of sound by the brain
- locating the source of sound
- testing for conductive and neural deafness.

Suggested reading

Wilson 1990, pp. 289–292
Hubbard & Mechan 1987, pp. 276–289
Rutishauser 1994, pp. 433–444
Jennett 1989, pp. 386–401
Guyton 1984, pp. 240–247
Marieb 1992, pp. 521–533 (as usual there are some excellent diagrams).

An additional subject you might like to study is the function of the inner ear related to maintaining balance. Guyton, Jennett and Marieb cover this fully in the above references; in Rutishauser you will need to turn to *Ch. 24*.

SIGHT

Again, the following topics for study are suggested, together with appropriate references:

- the nature of light
- bending of light (reflection and refraction)
- the structure of the eye:
  - the transparent structures (cornea, lens and the fluids)
  - the retina
  - the iris and pupil
- tears and their 'drainage system'
- the purpose of blinking
- focusing, and associated problems
- responses of the pupil to varying light
- simple optical tests
- retinal response to light – rods and cones, optic disc, optic nerve
- colour blindness
- visual pathway, visual field, and the perception of vision in the brain
- binocular vision.

Suggested reading

Wilson 1990, pp. 292–302
Rutishauser 1994, pp. 445–457
Hubbard & Mechan 1987, pp. 263–276
Guyton 1984, pp. 223–238
Jennett 1989, pp. 371–386
Marieb 1992, pp. 501–521.

Circulatory systems

Contents

(continued)

UNIT 3

- ■ **Respiration — the mechanics of breathing**
  Work sheet (3 to 5 hours)

- ■ **Respiration — gas exchange**
  Guided study (about 2 days)

- ■ **The respiratory system**
  Check quiz

■ An overview of the blood (work sheet)

Time for completion   About 3 to 4 hours

Overall aim   To achieve a basic understanding of the structure and function of blood – its cellular and fluid compartments.

Introduction   If you were to try to describe blood to a visitor from another planet, you might begin with a couple of very obvious statements: it's red, and it's a liquid. You might then add, it feels sticky if you touch it, and it's thicker than, say, water or even milk. It's warm too; blood has the same temperature as that of the body.

But what if blood is spilled? You might then want to tell your other-world visitor that blood congeals when away from the body. You would probably also add that it is dangerous to lose too much blood, that blood is vital to life.

Despite the simplistic nature of the above description, we have already raised a number of important issues. What is it that gives blood its colour, its stickiness, its thickness? Why does blood clot, and why is blood vital to life? These are some of the questions this Work Sheet sets out to tackle.

Background questions

1. What is the normal blood volume of an adult?

2. What is the approximate blood volume of a new baby? (You may like to consider the implications of your answer for, say, surgery on a baby, or an accident in which blood is lost.)

Imagine you have collected a specimen of blood in a test tube. After spinning in a centrifuge the blood has separated into what looks at first like two differently coloured parts, as in Figure 3.1.

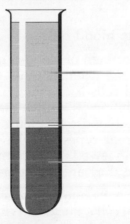

Fig. 3.1 A centrifuged specimen of blood.

3. Draw a diagram of such a test tube, labelling the two main divisions of the blood specimen.

4. Closer examination of the blood sample will reveal the presence of a third division, much smaller than the other two and separating them within the test tube. Label this third division. See. *Wilson 1990, Fig. 4.1A; Marieb 1992, Fig. 18.1; Rutishauser 1994, Fig. 4.7.*

Specific questions

BLOOD CELLS
Time for completion: about 1½ to 2 hours

The specimen's lower division in the test tube consists of solids, that is, blood cells. The great majority of these cells are red cells, and in health they form about 45% of total blood volume.

1. What is the term given to this cellular part of the blood volume? (You may find two terms in various texts, one of which can be abbreviated on blood result slips by its initials.) See *Rutishauser 1994, p. 61*.

Red blood cells

2. What is another name for the red blood cell?

3. Draw a simple diagram of a red cell, noting its distinctive features and size.

A patient suspected of being anaemic will have a blood test called a full blood count (FBC). One part of this test is an estimation of the number of red blood cells in a given volume of blood (red cell count: RCC or RBC).

4. What is the normal range of RBC for an adult? You will notice in your textbook that a different RBC is given for males and females. Make a note of both, and also find out why the RBC should differ between the sexes.
   You should note two things about the RBC. First, the norm is given as a range rather than a fixed number. Second, the RBC consists of a number for a given volume of blood. Without stating the volume of blood, any red cell count is, strictly speaking, useless.

5. State briefly the function of the red blood cell – a simple overview of one or two sentences will be sufficient here. A later Guided Study will consider the development of the red blood cell.

For Questions 1 to 5 see: *Wilson 1990, pp. 58–60; Rutishauser 1994, pp. 62–64*.

White blood cells

A specimen of blood can also be measured for its white cell count (WCC) and here the haematologist will differentiate between the various types of white blood cells.
   White cells, unlike mature red cells, have a nucleus. Another name for the white blood cell is leucocyte. Remember, this is the family name, covering all different types of white blood cells. Some leucocytes can be differentiated when viewed under a microscope by the presence of granules in the cell body: hence they are sometimes referred to as granulocytes.

6. They have another, much longer, name, deriving from the many-lobed shape of the cell's nucleus. What is this name? (You may also like to note its shorter form which you may hear used on the wards.)

7. Draw diagrams of the three types of granulocytes, naming them, and showing their nuclei and granules. See: *Wilson 1990, Fig. 4.5; Rutishauser 1994, Fig. 4.12*.

8. What is the normal range of white cell count (WCC)?

9. If you have a sore throat or an infected wound, your WCC may rise. With this in mind, state briefly (no more than a couple of sentences) the overall function of granulocytes. There is no need at the moment to state different functions for the three types of granulocytes.

10. What is the nature of the granules found in granulocytes?

For Questions 6 to 10 see: *Wilson 1990, pp. 51–52; Rutishauser 1994, pp. 66–67.*

Non-granular white blood cells

There are two broad types of white blood cells that have no granules in their cytoplasm. These are monocytes and lymphocytes. Their functions are related to those of the granulocytes though there are important differences.

11. Draw diagrams of a monocyte and a lymphocyte, noting the main differences in appearance between these agranulocytes (or non-granulocytes) and the granulocytes you drew earlier.

12. Try to derive from your chosen physiology textbook a generalised statement about their functions. Again, don't go beyond two or three sentences.

For Questions 11 to 12 see: *Wilson 1990, pp. 52–54; Rutishauser 1994, pp. 67–68.*

 FURTHER STUDY

You will undoubtedly have heard of AIDS (acquired immune deficiency syndrome) and may perhaps have read of the connection between this syndrome and lymphocytes. Can you recall from your general knowledge, such as newspapers or television programmes, how AIDS affects the sufferer? There are many diseases associated with AIDS, but try to produce an overview of the effect AIDS has on the human body. (See, for example, *Marieb 1992, pp. 716–717*, though you may question the author's use of the word 'plague'.)

Platelets

In *Fig. 4.3* of *Wilson 1990* you can see small blood cells containing no nucleus. These are sometimes called platelets.

13. What is an alternative name for platelets?

14. What is the normal platelet count, and what, very briefly, is the function of the blood's platelets?

15. Can you think of any disorder where this function is diminished?

For Questions 13 to 15 see: *Wilson 1990, pp. 60–61; Rutishauser 1994, pp. 68–69.*

 This is an appropriate place for a break.

THE PLASMA
Time for completion: 1¼ to 2 hours

Blood cells need a fluid medium in which to travel, and this fluid is called the plasma. Because of its fluidity, and because of the vast network of blood vessels which extends throughout the body, the blood is an excellent means of communication between parts of the body. We might call the blood, therefore, a 'transport system'. The main proportion of the plasma, about 90%, is water.

16. What percentage of the total blood volume is plasma? A long list of the constituents of plasma may be difficult to remember, so perhaps it would be useful to put these constituents into groups, such as the following:

- nutrients absorbed from the gut
- waste products derived from metabolic processes
- mineral salts derived from the diet
- hormones and enzymes manufactured by various glands
- dissolved gases from inspired air
- plasma proteins.

Of these, the plasma proteins are chosen for further consideration here. Some of them play a part in the blood clotting mechanism (which is considered in a later Work Sheet). Some are important in protecting the body against invading microorganisms (again, served by a more detailed Guided Study). It is the plasma proteins that give blood its 'stickiness' or viscosity, but they have other functions as well.

17. Which of the plasma proteins is present in the greatest quantities?

18. What part do the plasma proteins play in the transport of hormones around the body?

19. What is the connection between plasma proteins and the osmotic pressure of the blood?

20. What large organ in the body is important in the production of many of the plasma proteins?

21. Describe briefly the function of antibodies, and under what circumstances they can be produced.

For Questions 17 to 21 see: *Wilson 1990, pp. 49–50; Rutishauser 1994, pp. 69–70.*

THE pH OF BLOOD

Different parts of the body are either acid or alkaline. (As background information, you might like to recall which is the most acidic part of the body.) The amount of 'acidness' or acidity is expressed as a number: a highly acid substance might have a pH of 2.1, for example, and a highly alkaline substance might have a pH of 12.5. The term pH is a way of expressing the hydrogen ion concentration of the substance. For a fuller explanation see: *Wilson 1990 p. 44; Rutishauser 1994, p. 225.*

22. What is the normal pH of the blood, and what is its normal range?

23. Can you think of any commonly used foodstuffs, or medicines, that might affect the pH of the blood?

FURTHER STUDY

On completing this Work Sheet you will have gained an understanding of many of the entries on a blood test form (e.g. haemoglobin levels, RCC, WCC). With the assistance of your tutor or ward mentor, examine a results form from the haematology department and note how many other measurements are carried out. Look out in particular for the following:

- haematocrit or PCV
- mean cell volume (MCV)
- mean cell haemoglobin (MCH)
- mean cell haemoglobin concentration (MCHC).

Note the normal levels for these, and others you may find; but, more important, try to understand what these terms actually mean. By doing this, the mysteries of a patient's 'blood slip' will gradually become clear.

■ Blood cells — appearance and functions (guided study)

Time for completion   Allow between 1 and 2 'college days', with 5 to 6 hours' study per day

Overall aim   To develop the student's background knowledge of the blood, and to explore in greater detail the functions of red cells, white cells and platelets.

Introduction   Blood is vital to life. Insufficient red blood cells threatens the carriage of oxygen to tissue cells. Blood clotting is necessary to prevent loss of too much blood following injury, yet clots forming within blood vessels can cause serious ill health, even death. We rely on certain white blood cells to help protect us against harmful microorganisms. Certain cytotoxic drugs, used to fight cancer, can reduce the number of white blood cells, thus threatening the individual's ability to ward off infection.

In this Guided Study we examine the roles of the various blood cells. By understanding how these cells function in health, nurses can appreciate more fully what may happen when something goes wrong with a patient's blood cells. They will then be in a better position to explain treatments to their junior colleagues and to the patients themselves.

Background questions   Figure 3.2 shows a collection of blood cells as if viewed under a microscope. By chance, this microscope slide includes the main types of blood cells.

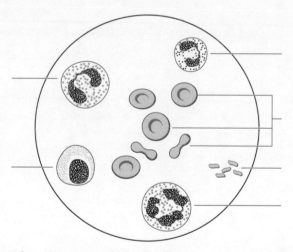

Fig. 3.2 Slide showing the main types of blood cell.

1. Copy Figure 3.2 and name the different blood cells shown, giving alternative names where appropriate. Check your diagram against *Fig. 4.3* of *Wilson 1990*. (Figures tend to be a lot clearer than 'the real thing', so for comparison look at the photomicrograph of blood cells in *Marieb 1992, Fig. 18.2*.)

2. State briefly the function of each type of cell you've drawn. (This should be revision following your completion of the earlier Work Sheet.)

Specific questions   **SECTION 1: ERYTHROCYTE (red blood cell)**
Time for completion: about 2 hours

You have already described the appearance and size of the erythrocyte, and noted the normal range of the red cell count (RCC or RBC) in both males and females. We now examine how the red cell develops and, in greater detail, its function, life span and destruction.

Erythropoiesis

1. Define the term erythropoiesis.

2. Where in the body are blood cells, including the erythrocytes, formed? You may like to show this in diagram form, as in *Marieb 1992, Figs 18.6 and 18.7*.

3. What factors are required for red cell production and development? Your answer should include substances taken in the diet that are necessary for the production and development of erythrocytes; and also a hormone produced by the kidneys that has a direct influence on the rate of production of red cells.

4. What other hormones influence erythropoiesis?

 COMMENTARY ON QUESTIONS 1 TO 4

Here it is a good idea to consult a number of texts of increasing complexity: *Wilson 1990, p. 59*, gives an adequate introduction from which you can make basic notes; then move on to *Marieb 1992, pp. 582–584*, and *Rutishauser 1994, p. 62*.
   Also worth consulting are *Guyton 1984, pp. 392–393*; and *Guyton 1991, pp 358–360*.

Stages of development

Red cells move through a number of stages of development. It is probably unnecessary to remember all the names of these stages (which tend to vary slightly from one text to another), but it is important to grasp the following points:

- Immature cells can appear in the bloodstream in certain medical conditions, as well as mature erythrocytes.
- As the red cell develops it loses its nucleus until it becomes no longer a true cell.
- As the red cell develops, it accumulates haemoglobin.

5. Draw a diagram of the stages of development showing these points. At what stage in the development process do mature red cells normally appear in the blood? See: *Rutishauser 1994, Fig. 4.8; Hubbard & Mechan 1987, Fig. 2.9; Marieb 1992, Fig. 18.5; Guyton 1991, Fig. 32.3.*

The mature erythrocyte — the function of haemoglobin

6. Describe the structure of haemoglobin, and show it in diagram form. What is the normal range of haemoglobin for men and women?

7. Describe the function of haemoglobin with relation to oxygen and carbon dioxide. Go into more detail than in the preceding Work Sheet, but bear in mind this topic will be covered further in a Guided Study on the respiratory system.

8. State what factors are important for the development and maintenance of haemoglobin in the red blood cell. See: *Rutishauser 1994, p. 62; Hubbard & Mechan 1987, pp. 74–75; Marieb 1992, pp. 580–584, including Fig. 18.4; Guyton 1991, pp. 358–361.*

   (If consulting American texts such as Marieb, note their spelling of hemoglobin.)

 ACTIVITY

Find out the recommended average daily intake of iron in the diet. Are there any groups of people who require a higher intake, and why?

You will have seen shelves of multivitamin tablets in chemists and supermarkets. Some of these contain iron, but in what quantity? How does this compare with the recommended daily intake?

Some iron tablets are available with a doctor's prescription (e.g. ferrous sulphate, and ferrous gluconate). Find out the usual daily dosage of these drugs. See, for example, Hopkins 1992.

Do multivitamin tablets contain substances other than iron that are necessary for red cell development?

With your colleagues plan a day's meals that would be suitable for an elderly person with mild iron-deficient anaemia. Your meals should be cheap, and prepared in such a way that the cooking process doesn't destroy important nutrients. Show which of the foodstuffs in the menu you are planning contain iron.

Life span of the erythrocyte

In health, new red cells are produced at the same rate as old red cells are destroyed.

9. Approximately how long does it take for a red cell to develop from a stem cell to a mature erythrocyte?

10. What is the normal life span of an erythrocyte?

11. How are erythrocytes broken down, and whereabouts in the body does this occur?

12. How is haemoglobin broken down, and what happens to its constituents?

 FURTHER STUDY

In some diseases, a blood test called erythrocyte sedimentation rate (ESR) is performed. Find out what this test is and how it is carried out. What is a normal ESR, and what is the significance of a raised result? What diseases can be tested for by the ESR?

What are the limitations of a raised ESR result?

See *Hinchliff & Montague 1988, p. 280.* (You could also consult a medical nursing text.)

 This would be a good place to have a break.

SECTION 2: LEUCOCYTES (white blood cells)
Time for completion: about 3 hours

You should have with you the completed initial Work Sheet of this Unit, which will act as a useful foundation for your further study.

Granulocytes (polymorphonuclear leucocytes)

You have already drawn a diagram showing the three types of granulocytes.

13. Name the three types of granulocytes (or polymorphs), and describe their functions more fully than the brief descriptions you've already given.

14. What are the normal ranges for these three types? Explain the staining system that gives granulocytes their colour as viewed through a microscope.

15. Define the terms leucopenia and leucocytosis. Try to work out under what medical conditions they might occur.

16. Name the stages of development of polymorphs, using a diagram to illustrate the process. (As with erythrocytes, it is probably unnecessary to memorise the names of all the stages the white cells pass through.)

17. Polymorphs have a number of properties, listed below. Define each of these terms, illustrating your definitions, where appropriate, with a diagram:

    - amoeboid movement
    - diapedesis
    - chemotaxis
    - phagocytosis (hence the family name phagocyte, sometimes used as an alternative to granulocyte)
    - digestion of phagocytosed particles.

For Question 13 to 17 see: *Wilson 1990, pp. 51–52, and Figs 4.5, 4.6, and 4.7; Rutishauser 1994, pp. 67–68 and 282–284; Marieb 1992, pp. 587–591, and Fig. 18.9; Guyton 1984, pp. 396–399; Guyton 1991, pp. 365–368.* (The figures in both books by Guyton are less clear than that in Marieb.)

Agranulocytes

Two main types of white cell are without granules: lymphocytes and monocytes. Both play an important part in the body's defences against infection.

Lymphocytes

18. Describe the formation, development and life cycle of lymphocytes, differentiating between the B lymphocyte and the T lymphocyte.
    As before, you may wish to begin with the fairly basic description given by *Wilson 1990, pp. 53–54,* then move on to: *Rutishauser 1994, pp. 67–68 and 284; Hinchliff & Montague 1988, pp. 287–288; Guyton 1984, pp. 407–408.*
    The function of lymphocytes is dealt with later.

Monocytes

19. Describe how monocytes are formed and developed, and state their functions.

20. Where are monocytes found in the body, and what other names are they given?

21. Agranulocytes help protect the body against infection by attacking bacteria and viruses. What other 'invaders' can they attack and destroy?

This is an appropriate place for a break.

SECTION 3: IMMUNITY
Time for completion: about 3 hours – for example 1 morning

In this part of the Guided Study, we consider the part played by white cells in the defence of the body against invasion by foreign particles including microorganisms. You've already described actions, such as diapedesis and phagocytosis, in which bacteria are engulfed and digested by granulocytes. An end-product of this process is pus (see *Wilson 1990, p. 52*; and *Rutishauser 1994, p. 67*).

Now we turn to the protection given by the two types of lymphocytes – T cells and B cells. This form of defence is often described as specific immunity, because the protection that develops against, say, measles doesn't provide protection against any other 'invader'; it is specific to measles.

Role of the B lymphocytes

Let's assume that bacteria have gained access to a person's tissues through a cut in the skin, and now they are multiplying.

22. First, discover what the 'B' in B cells stands for. (This isn't vital for your understanding of immunity, but you might be curious about the name.)

23. Now read, and make notes on, how invading bacteria (which we'll refer to as an antigen) activate B cells to divide. What is an antigen? What makes an invading substance antigenic? (For example, if a pollen grain that gains access to the body is an antigen, is a transplanted kidney also an antigen? Is a metal hip joint an antigen? What is it that makes antigens 'foreign'?)

24. Find out what plasma cells are, and what antibodies are. Note that there are different types of antibodies, with differing molecular structures.

25. What are memory cells?

26. Distinguish between the primary response and secondary response to invasion of the body by antigens.

 COMMENTARY ON QUESTIONS 22 TO 26

For useful reading around this subject, see *Guyton 1984, pp. 409–410*. There is a brief mention here of the complement system which you may care to note, but don't get bogged down with it: it's extremely complex. *Marieb 1992*, goes into much greater detail about antibody production, plasma and memory cells, and the primary and secondary responses. There are some excellent diagrams; see *pp. 697–708*. Fig. 22.9 on p. 702 gives a helpful overview of the role of the B lymphocyte. See also *Rutishauser 1994, Chs 4 and 15*.

Role of the T lymphocytes

T cells do not manufacture antibodies, so they are not part of the body's humoral response. (Incidentally, what does humoral mean?) They attack invading antigens directly (killer or cytotoxic cells). Some T cells, however, act as helper and suppressor cells.

27. What does the 'T' stand for, in T Lymphocyte? Describe the production and functions of each of these types of T cells:

    - killer cells
    - helper cells
    - suppressor cells.

See: *Rutishauser 1994, pp. 286–288; Marieb 1992, pp. 708–712.*

 COMMENTARY ON QUESTION 27

One way of checking your understanding of this difficult topic is to draw a flow chart representing the passage of an antigen through the various body defence mechanisms, and illustrating the functions of polymorphs, B cells and antibodies, T cells and monocytes. This will also illustrate how specific and non-specific defences, and humoral and cellular responses, all work together to protect the individual.

Figure 3.3 provides you with the beginning of such a flow chart.

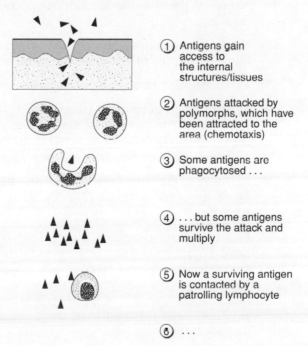

1. Antigens gain access to the internal structures/tissues

2. Antigens attacked by polymorphs, which have been attracted to the area (chemotaxis)

3. Some antigens are phagocytosed . . .

4. . . . but some antigens survive the attack and multiply

5. Now a surviving antigen is contacted by a patrolling lymphocyte

6. . . .

Fig. 3.3 First stages of the body's defences against infection. Complete the diagram.

FURTHER STUDY

Related topics for further study could include:

- AIDS and other immunodeficient conditions
- autoimmune conditions (check your understanding of this term)
- transplanted organ rejection
- allergic response
- active and passive immunity, and the role of vaccines.

 Take another break here.

SECTION 4: THROMBOCYTES (platelets)
Time for completion: about 2 hours

In the preceding Work Sheet you noted the normal range of platelets for an adult, and gave a brief description of their function. The remainder of this Guided Study deals with the important subject of blood clotting.

You know from experience that a small cut will bleed for a little while, but after a few minutes, perhaps with the application of pressure, the bleeding stops. Blood clotting is an important part of the body's defence against injury, since continued bleeding would have serious consequences. You may have heard of the disease haemophilia where one factor in the clotting sequence is missing.

28. What are the consequences for a person with haemophilia following trauma? Give examples of some of the clinical features of this condition.

29. What regular treatment is required by a haemophiliac?

30. Write down, initially, a simple overview of the clotting process: what triggers it off, and the cascade shape of the clotting sequence (but don't get bogged down by long lists of names of factors). What part do platelets play in the clotting process? See, for example, *Rutishauser 1994, p. 70.*

31. Name the clotting factors that are present in the blood in an inactive form. How are they activated? Why do you think they exist initially in inactive form?

32. Draw a flow diagram illustrating the sequence of events from the initial tissue damage (e.g. a cut finger) to the formation of a blood clot. Describe the structure of a blood clot.

Guyton 1984, pp. 421–422, describes two mechanisms for initiating the clotting process – the intrinsic and the extrinsic. See also *Rutishauser 1994, pp. 72–73, and Fig. 4.16.*

33. Note what sets off these mechanisms, how they differ from each other, and where they merge into the formation of the blood clot itself.

34. Note the place of factor VIII in the clotting sequence (of importance because of its role in the disease haemophilia).

35. Describe the role of the liver in the production of certain clotting factors.

36. There are many blood tests associated with blood clotting. Define the following:

- bleeding time
- clotting time
- prothrombin time.

For Questions 30 to 36 see: *Rutishauser 1994, pp. 71–73; Marieb 1992, pp. 592–597; Hinchliff & Montague 1988, pp. 289–297* (provides far more detail, so get a broad understanding first).

 FURTHER STUDY

Think about the effect on a person's clotting mechanisms of liver disease. What drugs can be given to increase clotting time, and when might such drugs be given? What observations should the nurse make on a patient being given such drugs?

REFERENCE
Hopkins S J 1992 Drugs and pharmacology for nurses, 11th edn. Churchill Livingstone, Edinburgh

■ Blood groups (work sheet)

Time for completion
About 3 hours

Overall aim
To develop the student's knowledge of the ABO and Rhesus systems of blood grouping, and to explore how donated blood is cross-matched with that of a recipient.

Introduction
Blood transfusions can save a patient's life. Severe trauma, as in a road accident, can lead to massive loss of blood. Major surgery almost always leads to blood loss, although techniques for controlling this loss during operation are constantly being improved.

When a patient is first admitted to Accident Service, and requires a blood transfusion, there are certain intravenous fluids that can be given initially (such as plasma, and plasma expanders) but the haematology laboratory will need to cross-match sufficient units of blood for transfusion into the patient. Blood is required because of its oxygen-carrying properties. If the patient's blood volume falls, there is reduced oxygen supply to vital areas like the brain and kidneys.

Some of you may be blood donors, because in the UK donating blood is a voluntary service. Blood donors will be aware, from the certificates they are awarded, that they belong to a particular blood group: O, A, AB, etc., and that some groups are rarer than others. They will know that patients can be given only blood that is compatible with their own, and that serious consequences may arise if incompatible blood is given by mistake.

In this Work Sheet, we look at the meaning of the term blood group, and how the laboratory works out which group a unit of donated blood belongs to.

Background questions
You may hear people talk about donating a pint of blood, or being given 2 pints of blood when they were in hospital. Blood isn't nowadays measured in pints.

1. Find out the volume of blood taken from a donor at each session.

2. Read about how donated blood is treated and stored. What chemicals are added to it and why? What safeguards are there to prevent the passing on of diseases like hepatitis and syphilis? The handbook of the United Kingdom Blood Transfusion Services (1989) is a useful source of information.

Specific questions
THE ABO SYSTEM
Time for completion: about 1 or 1½ hours

1. Certain antigens are found on the surface of red blood cells. Check that you understand the meaning of antigen. It has occurred in an earlier Guided Study. The red cell surface antigens are sometimes called agglutinogens, but for the moment we'll use the shorter term.

   A person's blood group is determined by the presence or absence of these inherited antigens on the red cell surface. The antigens in the ABO system are called antigen A and antigen B.

2. Look at Figure 3.4 (a highly figurative diagram) and, with reference to your chosen physiology text, name the blood group to which each red cell belongs. See: *Wilson 1990, p. 60; Rutishauser 1994, pp. 64–66;* or *Guyton 1984, p. 426.*

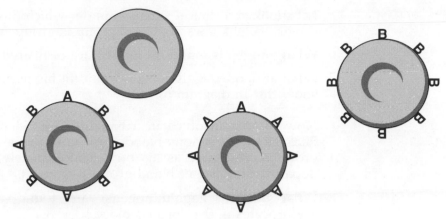

Fig. 3.4 Antigens on the surface of red blood cells.

You will see from this diagram that a person's red cells can have:

- A antigens, or
- B antigens, or
- both A and B antigens, or
- neither A nor B antigens.

 FURTHER STUDY

At a later stage you may wish to study the subject of blood groups more deeply; you are recommended to read *Hinchliff & Montague 1988, pp. 300–304*. You'll see there that there are many more antigens than A and B. At the moment, however, let's keep to just A and B in order to grasp the basic principle of red cell antigens.

So far, then, we have a blood group which is genetically determined by antigens (or their absence) on our red cell membranes. These antigens develop within the fetus, and are present at birth.

3. In the plasma, there are antibodies present. Again, check up on your understanding of this term, which, like antigen, came up in an earlier Guided Study. These antibodies are also called agglutinins, and are named after the antigen (agglutinogen) with which they react.
   Thus:

   - blood group A has cells containing antigen A, and its plasma contains antibody anti-B
   - blood group B has cells containing antigen B, and its plasma contains antibody anti-A.

4. What antibodies are present in the plasma of blood group O and blood group AB?

5. If antigen A meets antibody anti-A, or if antigen B meets antibody anti-B, the red cells haemolyse and agglutinate. Find out what these terms mean. See *Rutishauser 1994, Ch. 4*, 'Transfusion reactions'.

Let's look at a clinical situation under which this might occur. John is a blood donor, so he knows his blood group is group A.

6. What antigen is found on the surface of his red blood cells?

7. What antibody (agglutinin) is found in his plasma? (You may find it helps to show this in diagram form.)

Now John is involved in a bad traffic accident and is taken into hospital. Because of his serious blood loss, he requires a blood transfusion of blood group A, the same as his own. Unfortunately, in the confusion of a busy department, a bag of blood group B is given.

8. What antigens (agglutinogens) are found on the cell membranes of erythrocytes in the donated blood?

9. What physiological events will happen when those donated red cells, transfused into John, meet up with the antibodies (agglutinins) you have identified as being present in his plasma?

 COMMENTARY ON QUESTIONS 6 TO 9

Remember that it is the donated cells that are damaged (haemolysed) by the agglutinins in the recipient's plasma. (*See Wilson 1990, p. 60.*) This distinction is made more fully by *Hinchliff & Montague 1988, p. 304.*

 FURTHER STUDY

While this Workbook is concerned with physiology rather than disease (pathology) it would be useful for you to discover the clinical features of a patient who is receiving an incompatible blood transfusion. Note also the emergency response by nursing staff when incompatibility is suspected. Remember that an incompatible transfusion, if allowed to continue, could kill your patient.

What is the best way of avoiding the error that happened with John's transfusion?

CROSSING-MATCHING OF BLOOD
Time for completion: ¾ to 1 hour

You already know that there are many more antigens than the A and B antigens mentioned in this Work Sheet. In practice, then, to avoid incompatibility arising via these rarer antigens, donated blood is cross-matched with a specimen of blood taken from a patient who requires a transfusion. In this way the laboratory ensures that particular units of donated blood are indeed compatible with the blood of a particular patient.

10. Find out what information and instructions are included on the labels of units of blood used in your hospital. Read your hospital's policy on blood transfusion.

11. Cross-matching is carried out in the hospital's haematology department.

Make notes on how the procedure is performed. See, for example, *Marieb 1992, pp. 597–600*, for a very full discussion.

12. What is the usual time taken for a unit of donated blood to be cross-matched ready for transfusion? How long would this process take in an emergency?

THE RHESUS FACTOR
Time for completion: ¾ to 1 hour

As well as the ABO blood group system, there is another major classification of blood groups, called the Rhesus system. The majority of people are born with the Rhesus factor present on their red cell membranes, and so are referred to as Rhesus positive (Rh+). Consequently a person will be A Rh+, or O Rh+, or AB Rh+ etc. People without this Rhesus factor are Rhesus negative (Rh–), and so they belong to blood groups such as A Rh–, O Rh–, etc.

There are several Rhesus factors of which the commonest is factor D. (See: *Wilson 1990, p. 60*; and in greater detail *Hinchliff & Montague 1988, pp. 303–305*.)

13. People who are Rh+ do not have the Rhesus antibody in their plasma. State why not. However, people who are Rh– also do not *naturally* have the Rhesus antibody in their plasma, but this antibody can develop under certain circumstances.

 COMMENTARY ON QUESTION 13

Here is one major difference between the Rhesus system and the ABO system. People who are blood group A, for example, have naturally occurring antibody B in their plasma. People who are blood group B have naturally occurring antibody A in their plasma. Rhesus negative people, however, do not have naturally occurring anti-Rhesus antibody in their plasma. It has to be caused to develop.

14. Describe how these antibodies might develop. You will need to discover under what circumstances Rhesus antigens come to be introduced into the blood of of a Rhesus negative person.

 COMMENTARY ON QUESTION 14

You may find it easier to deal first with a Rhesus negative patient being mistakenly given a transfusion of Rhesus positive blood. How long will it take for that patient to develop antibodies to the Rhesus antigen?

Then go on to the more complex example of parents of mixed Rhesus groups conceiving a Rhesus negative baby. You may find it clearer to draw a series of diagrams. Have a look at *Bloom & Bloom 1986, p. 231, Fig. 10.10.* This textbook of medical nursing contains some excellent diagrams which help with our understanding of many complex conditions.

FURTHER STUDY

It is worth reading about how blood groups are inherited. Blood tests showing blood groups, incidentally, don't prove parentage (in the case, for example, of a paternity claim in a court), but it is possible for a blood test to show that the accused man cannot be the father of a child. See *Guyton 1991, pp. 385–386.*

REFERENCES

Bloom A, Bloom S 1986 Toohey's medicine for nurses, 14th edn. Churchill Livingstone, Edinburgh

United Kingdom Blood Transfusion Services 1989 Handbook of transfusion medicine. HMSO, London

■ The blood (check quiz)

1. A junior nurse is reading the blood results slip of one of his patients, and he asks you to explain some of the results.

   a. State a haemoglobin level that would be normal for an adult male. Don't forget to match the figure you give with the correct unit.

   b. State a red cell count (RCC or RBC) that would be normal for an adult male.

   c. Explain to the student what is meant by haematocrit, and under what medical or physiological circumstances this might rise or fall.

2. Describe the importance of dietary iron for the development of red blood cells. Give five examples of fairly inexpensive foods containing substantial amounts of iron. How much iron do we need, on average, in our daily diet?

3. One of your patients has developed a urinary tract infection following catheterisation. How might this infection affect the patient's white cell count (WCC) and why?

4. a. Explain the differences in appearance and function between granulocytes and lymphocytes.

   b. In simple terms, explain to a fellow student the main differences between B lymphocytes and T lymphocytes.

   c. What changes might you expect on the blood results slip of a young patient with HIV? (This is a complex question, and there is probably no one correct answer, but be prepared to back up your answers with physiological reasons. It would be a good idea to discuss this question with a group of your colleagues.)

5. Your mentor has asked you to collect a bag of blood from the hospital's 'blood fridge' for transfusion into one of your patients.

   a. How would you check the blood for suitability for transfusion?

   b. What observations should you make on the patient while she is receiving the blood, and why? (You can attempt this part of the question if you tried the Further Study on page 94 of this Unit.)

6. You are being shown round the haematology laboratory where blood for transfusion is being grouped in case it is needed later. Figure 3.5 shows three blood specimens being tested against serum from known blood groups. To which blood groups do each of the three specimens belong?

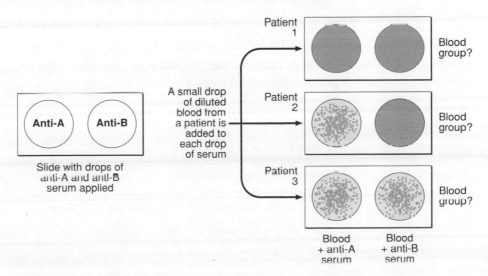

Fig. 3.5 Testing for blood groups.

■ An overview of the heart (work sheet)

Time for completion   About 3 or 4 hours

Overall aim   To achieve a basic understanding of the position of the heart within the thoracic cavity, its structure, and the main blood vessels leaving and supplying it.

Introduction   It is claimed that the healthy heart will beat between 1000 and 2000 million times during a person's life (*Jennett 1989, p. 151*). The heart's function is to supply the body with blood rich in oxygen, and to clear away waste products of various organs. If the blood supply to any organ ceases, that organ will die. (The brain is particularly susceptible to oxygen deprivation.) The heart can adapt, in health, to many differing demands and conditions of the body, from sleep to intense bursts of exercise as in sprinting, from deep-sea diving to high altitude mountaineering.

Background questions

1. State the position of the heart in the thoracic cavity, in relation to the lungs, trachea, diaphragm and oesophagus. Refer to your preferred physiology text (for example *Rutishauser 1994, p. 82; Wilson 1990 pp. 73–74*).

2. Briefly give a description of the size and shape of the adult heart. Make sure you incorporate the terms apex and base.

3. With the aid of a diagram, show how the ribs and sternum partially surround and protect the heart. Where is the apex of the heart in relation to the ribs? Much of the heart lies under the sternum: what might be the implications of this for nursing practice? (Think about how cardiac massage is performed.)

Specific questions

STRUCTURE OF THE HEART
Time for completion: about 1 hour

There are three layers of tissue that make up the heart.

Pericardium — the outermost layer

1. State the type of tissue that forms the pericardium.

2. Describe how the pericardium is connected to the diaphragm. What purpose might this have?

3. What is the meaning of the term potential space and how does it relate to the pericardium? (Where else in the body is a potential space found?) Describe the function of serous fluid found between the pericardial layers.

Myocardium — the middle layer

4. Myocardium is highly specialised muscle, the tissue that actually causes the heart to beat. Refer to the notes and diagrams you made on cardiac muscle in Unit 1 of this Workbook. Or draw a diagram of this muscle tissue, and attempt to show how it differs in structure and function from skeletal muscle. Refer, for example, to: *Rutishauser 1994, p. 389; Mackenna & Callander 1990, pp. 16 and 92*.

5. Note where in the heart wall the myocardium is thickest. Draw a simple diagram showing this. What do you think is the reason for this additional thickness of muscle?

Endocardium — the inner layer

6. Describe the structure of the endocardium and note its relationship with the lining of adjacent blood vessels and the covering of heart valves.

INTERNAL STRUCTURE OF THE HEART
Time for completion: about half an hour

Figure 3.6 is a highly simplified diagram showing the chambers of the heart.

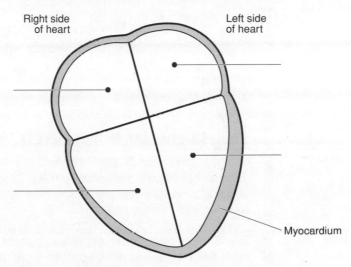

Fig. 3.6 The chambers of the heart.

7. Name the four chambers of the heart shown in Figure 3.6. Remember that the diagram shows the heart in its anatomical position (with its left side on the right side of the page).

8. State what structures divide the heart into both left and right, and upper and lower parts. Show these divisions on your own diagram.

9. Redraw the diagram (but retaining its simplicity) and show the heart valves:

   • between the upper and lower chambers on both left and right sides of the heart
   • between the heart's lower chambers and the major blood vessels.

In a later Guided Study we shall consider the flow of blood through the heart.

THE MAJOR BLOOD VESSELS
Time for completion: about half an hour

The heart pumps blood to the lungs from the right ventricle.

10. Add to your diagram the arteries that carry blood from the heart to the lungs, and name them. Note whether they convey oxygenated or de-oxygenated blood (perhaps using the colours red and blue to differentiate the two).

The heart also pumps blood to the rest of the body from the left ventricle via a major artery.

11. Show this artery on your diagram, and name it. Does it convey oxygenated or de-oxygenated blood? You may like to investigate how far this artery travels

through the body, and how its name changes slightly as it passes through different parts of the body.

As well as pumping out blood, the heart also receives blood, both from the lungs and from the remainder of the body.

12. Which major blood vessels return blood from the left and right lungs to the heart? Show them on your diagram, and demonstrate to which heart chamber blood is returned. Are these major blood vessels called veins or arteries; and do they convey oxygen-rich or oxygen-poor blood?

13. Which major blood vessels (veins or arteries) return blood to the heart from all of the body except the lungs? Which chamber do they return it to? Add these vessels to your diagram, and use your chosen colour code to demonstrate whether the blood conveyed is high or low in oxygen.

 COMMENTARY ON QUESTIONS 10 TO 13

You may feel your diagram, showing the heart, its chambers and the major blood vessels, has become somewhat complex. Check it against that in *Wilson 1990, Fig. 5.8*, which also usefully demonstrates, by use of colour, whether the blood being carried is oxygenated or de-oxygenated.

You will know that arterial blood is *usually* oxygenated, and that venous blood is *usually* de-oxygenated (having given up some of its oxygen to the body tissues). In the heart and its major blood vessels, however, this is not always the case, as you've just discovered.

THE BLOOD SUPPLY TO THE HEART TISSUES
Time for completion: about half an hour

You have now discovered which major blood vessel takes blood from the heart to much of the body. This is so that the tissues and organs of the body can be supplied with nutrients and oxygen, on which they depend for their survival.

The heart tissues themselves also require oxygen and nutrients. Considering that the heart has to beat continuously throughout a person's life, it is reasonable to assume that the heart tissues (especially the myocardium) require a good supply of oxygen.

14. Draw a diagram of the outside of the heart, and show the left and right coronary arteries and their major branches. Refer, for example, to: *Wilson 1990, Fig. 5.13; Hubbard & Mechan 1987, Fig. 4.14; Rutishauser 1994, Fig. 5.15.*

15. Show where the coronary arteries leave the aorta.

16. State briefly what is meant by the term cardiac output. What percentage of the cardiac output goes to the coronary arteries?

17. How is blood returned from supplying the heart (having given up some of its oxygen) to the circulation? See: *Jennett 1989, pp. 169–171.*

NERVE SUPPLY TO THE HEART
Time for completion: about half an hour

We are well aware that if we are frightened, or are running to catch a train, the heart beat speeds up. Sometimes we can feel our heart pounding, as it beats strongly and quickly in order to push oxygenated blood around the body.

ACTIVITY

Ask a friend to record your pulse (over a full minute) after you have done some exercises like running up stairs. How does this compare with your normal pulse, for example while you are sitting quietly reading?

18. Describe the nerve supply to the heart which can cause the heart beat to speed up and to slow down. Restrict your answer, at the moment, to the sympathetic and parasympathetic nerve supply to the heart. Which of these acts as the 'accelerator' to the heart, and which acts as the 'brake'? See *Mackenna & Callander 1990, p. 99;* and read any relevant notes you made in Unit 2.

COMMENTARY ON QUESTIONS 18

The heart rate is influenced not just by nerve input but by circulating chemicals in the blood (such as noradrenaline). The conducting system of the heart will be covered in a later Guided Study, and there it will be discovered that the heart muscle has its own, intrinsic contractility. At present simply be aware that the heart rate (and the force with which the myocardium contracts) can be influenced by the autonomic nervous system.

■ The cardiac cycle (guided study)

Time for completion   About 2 'college days', with 5 to 6 hours' study per day

Overall aim   To develop the student's background knowledge of the heart's structure, by exploring the electrical activity of the heart, the action of the valves during the cardiac cycle, and the recording of this activity via the electrocardiogram.

Introduction   Before tackling this Guided Study, you should ensure that you have a good grasp of the information derived from the previous Work Sheet on the heart. You should be clear about the names of the heart's chambers, its valves, and the major blood vessels supplying and leaving it.

Once you understand the functions of the healthy heart, you will be in a good position to appreciate the events that may occur in ischaemic heart disease (such as myocardial infarction), one of the major killers in the Western world.

Like previous Guided Studies, this one is divided into several sections, and you should take a break between each. Before you move on to a new section though, read through the notes you've made on the previous one to make sure you fully understand the information you derived from it.

Background questions   1. Look at the diagrams you made when studying the previous Work Sheet, and check your knowledge of the heart's anatomy.

2. Revise your notes on the special nature of cardiac muscle (myocardium). See, for example: *Marieb 1992, p. 614; Rutishauser 1994, p. 389.*

Specific questions   ## SECTION 1: THE CONDUCTING SYSTEM OF THE HEART
Time for completion: Sections 1 and 2 are designed to take 1 whole 'college day'. Section 2 will take a little longer than Section 1 to complete.

We know that, in health, the heart beats continuously, its rate varying according to the level of activity of the body. For example, as we run a cross-country race or climb a mountain, our heart rate increases. As we relax in front of the television, our heart rate slows. How is this achieved? What causes the heart to beat so rhythmically, and to respond so appropriately to the demands of the body for oxygen?

1. Draw in your notes Figure 3.7 which shows the four chambers of the heart and two of the valves. Label these features.

Fig. 3.7 Simplified diagram of the heart and its chambers.

2. Now, with reference to your physiology text(s), add to your diagram the sino-atrial node (SA node) and make brief notes on its nature and function. Of what sort of tissue is it made?

3. Next add the atrio-ventricular node (AV node) again making notes about its nature and functions.

4. In some texts, the sino-atrial node is referred to as the pacemaker of the heart. Don't get this term confused with the artificial pacemakers which some patients have fitted, but explain why the term pacemaker is applied to the sino-atrial node.

   What happens to the atria as the electrical activity spreads across them from the SA node?

5. Explain how the electrical activity initiated by the SA node reaches the AV node, and how it then progresses to the ventricles. How is it that the ventricles contract, as a result of this electrical activity, a split second after the atria?

For Questions 1 to 5 see: *Wilson 1990, p. 77 and Fig. 5, 14; Hubbard & Mechan 1987, pp. 117–118 and Fig. 4.4; Rutishauser 1994, pp. 84–85 and Fig. 5.10; Jennett 1989, pp. 154–156 and Fig. 6.3; Hinchliff & Montague 1988, p. 331 and Fig. 4.2.17; Guyton 1984, pp. 260–262; Guyton 1991 pp. 111–114.*

 COMMENTARY ON QUESTIONS 1 TO 5

Your diagram should contain the following features:

- SA node
- AV node
- atrio-ventricular bundle (bundle of His)
- Purkinje fibres.

Your notes should give details of a succession of events, both electrical and muscular, starting from the SA node initiating an electrical impulse. Your notes should explain how the ring of non-conducting tissue separating the atria from the ventricles (the ring which contains the A-V valves) slightly delays the electrical impulse; and how this impulse manages to get through to the ventricles.

 Take a break here. You should aim to complete Section 2 by the end of this first 'college day' devoted to this Guided Study.

SECTION 2: BLOOD FLOW THROUGH THE HEART

The heart is divided into left and right sides by the septum to form a double pump. The right side receives de-oxygenated blood from the body and sends it to the lungs to pick up oxygen and give off carbon dioxide. The left side receives oxygen-rich blood from the lungs and pumps it round the body, including the heart's own muscle via the coronary arteries.

Though divided into left and right sides, in health both atria contract almost together, followed a split second later by both ventricles contracting together. This is achieved by means of the electrical activity you have already described.

Steady, rhythmical contractions of the atria and ventricles allow these chambers to fill adequately with blood, which is then pumped out again. The cardiac cycle is the succession of events that leads to one contraction, the pumping of blood out to the lungs and around the rest of the body. Usually each cardiac cycle results in one pulse felt, for example, at the wrist.

After working through this Section, you will have produced a series of small, simple diagrams of the heart showing each stage of the cardiac cycle. It may help your understanding of events if these diagrams can be placed in line across the same page of your notes.

6. Begin by copying Figure 3.8, and complete the labelling.

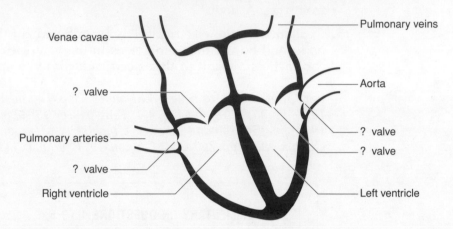

Venae cavae — Pulmonary veins
? valve — Aorta
Pulmonary arteries — ? valve
? valve — ? valve
Right ventricle — Left ventricle

Fig. 3.8 Simplified diagram of the heart and major blood vessels.

It is usual to begin describing the cardiac cycle at the stage when the atria begin to fill with blood (though you could start at any stage of the cycle).

7. Make a small-scale version of Figure 3.8 on the left of your page. Then, using arrows and shading, show blood entering both left and right atria, and remind yourself which vessels convey this blood. Show in your diagram the position of:

  • the valves between atria and ventricles
  • the valves between ventricles and arteries

demonstrating whether they are closed or open. Refer to your chosen text for assistance; for example *Mackenna & Callander 1990, p. 95.*

To help you begin your series of diagrams, Figure 3.9 is my version of

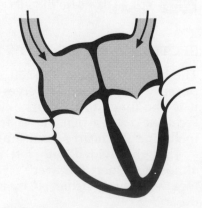

Fig. 3.9 Both atria fill with blood.

events at the beginning of the cardiac cycle. You'll see that I have shown all the heart valves closed.

8. As the atria fill with blood, the tricuspid and bicuspid (mitral) valves open, to let blood flow through into the ventricles.

    Show this on a new small-scale diagram, and explain in your notes accompanying it what it is that causes the valves to open. Again to help you initially I've drawn my own version of this part of the cardiac cycle (Fig. 3.10).

    What are the positions of the pulmonary and aortic valves during ventricular filling?

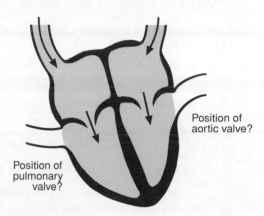

Fig. 3.10 Both ventricles fill with blood.

9. Note that while the ventricles fill with blood there is, at first, no contraction of the heart muscle. The myocardium is resting. What is the special name given to this period of the cardiac cycle? How long is it said to last if the heart rate is 72 beats per minute? See, for example: *Hubbard & Mechan 1987, p. 116; Wilson 1990, pp. 77–78; Rutishauser 1994, Fig. 5.13; Hinchliff & Montague 1988, p. 346.*

 COMMENTARY ON QUESTIONS 6 TO 9

At the moment, you are concentrating on the contraction of the heart chambers and the flow of blood through the heart. You may notice that some texts show diagrams that include a lot of information in a combined form: state of heart muscle; electrocardiogram; heart sounds; pressures within the heart chambers; and so on.

Such diagrams, while being informative, can initially appear confusing because there is too much information to take in. This is why I am trying to help you build up your knowledge – both in notes and diagram form – little by little. Eventually you may be able to produce your own complex diagram, but at a stage in your studies when it will be truly meaningful.

10. Now ventricular filling is completed by the atria contracting and forcing more blood into the ventricles – 'topping up' the ventricles. Again, show this stage of the cardiac cycle in a new diagram, showing the position of all four heart valves. Those between the atria and the ventricles (the A-V valves) will be open to allow the blood to flow through, but what about the pulmonary and aortic valves?

11. You have already discovered the name for the period of resting of the heart muscle; what name is given to the period when the ventricular myocardium contracts?

12. Show on another diagram the continuation of this stage of the cardiac cycle, when both ventricles contract. Use arrows to show the direction of blood flow, and make sure you demonstrate clearly the state of all of the valves. One pair of valves will be closed, and the other pair open – but which? It may help you to work this out to think where the blood is flowing to from the ventricles as they contract.
    What is it that actually causes the heart valves to open and close?

13. Once ventricular contraction has ended, and the blood in the ventricles has been expelled, the heart returns to its resting state, and the cardiac cycle begins again with the atria filling with blood. What is the position of each of the valves now?

 COMMENTARY ON QUESTIONS 6 TO 13

(Read this after you have attempted the questions.)

You should now have a series of simplified diagrams of the heart, showing the chambers and valves during different stages of the cardiac cycle. They will show the heart in its resting state (diastole) and its period of contraction (systole). Remember that systole applies to ventricular contraction. The atria contract during (i.e. right at the end of) diastole.

You will have noted the position of the heart valves, whether they are open or closed, during each stage. And you will have begun to understand why they open and close as they do – because of changes in pressure within the various heart chambers. To begin with you won't have put figures against the various pressures. But you will have appreciated, for example, that as the ventricles contract, pressure within them rises (because the blood contained in the ventricles is being squeezed by the heart muscle). It is this rise in pressure that shuts the A-V valves and opens the aortic and pulmonary valves.

Similarly, when the atria fill with blood (following atrial contraction) the rising pressure within the atria eventually outweighs pressure in the ventricles, which have now ceased contracting, and the A-V valves open.

Check your series of diagrams with, for example, those in *Mackenna & Callander 1990, p. 95*, and *Rutishauser 1994 Fig. 5.12*; and ensure that you fully understand the sequence of events described both here and in your notes.

 This is an appropriate place for a break. Sections 1 and 2 of this Guided Study will have taken you a whole 'college day'.

SECTION 3: THE ELECTROCARDIOGRAM

Time for completion: Sections 3 and 4 are designed to take a second 'college day'. Allow 3 to 4 hours for this Section.

When heart disease is suspected, the electrocardiogram (ECG) is an important diagnostic tool. It can show the rate and rhythm of the patient's heart beat, and also identify any area of damaged heart tissue, such as occurs following a myocardial infarction (heart attack).

The electrical activity generated by the myocardium during both systole and diastole passes from the heart itself into the surrounding tissues. A tiny electrical current actually reaches the surface of the skin where it can be picked up, and the size of the current measured, by the electrocardiograph. The resulting ECG can be displayed on a screen or printed out on a strip of graph paper.

In health a person's ECG has a typical shape (Fig. 3.11). You will see that various features on the ECG are labelled with either a group of letters (the QRS complex) or single letters (P wave and T wave).

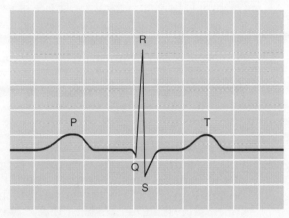

Fig. 3.11 A normal ECG over one complete cardiac cycle.

14. With reference to your preferred text, make notes on the activity of the heart muscle that coincides with these different parts of the ECG: the P wave, the QRS complex, and the T wave. See: *Rutishauser 1994, Fig. 5.11; Mackenna & Callander 1990, p. 98; Marieb 1992, pp. 619–622.*

15. Have a close look at a normal QRS complex, and copy it into your notes. Notice that it is, in health, a narrower wave than the P or T waves. The Q wave is a short downwards stroke, the R wave a much longer upward stroke, and the S wave downwards again. In myocardial infarction the QRS complex can alter in shape. (See *Rutishauser 1994, Fig. 5.11B.*) You may like to copy this altered ECG, too, for comparison with the normal ECG.

 COMMENTARY ON QUESTIONS 14 TO 15

Be careful how, in your notes, you express the significance of the different parts of the ECG. For example, it's all too easy to write that the P wave is caused by atrial contraction. This isn't the case. Atrial contraction actually begins at the peak of the P wave. It is the wave of excitation spreading across the atria from SA node which shows on the electrocardiogram and which itself brings about atrial contraction.

16. Some texts refer to the P wave as representing the depolarisation of the atria, and the QRS complex as representing the depolarisation of the ventricles (*Hinchliff & Montague 1988, p. 331*). Find out what the term depolarisation means.

What does the T wave represent?

FURTHER STUDY

Changes in the heart's rhythm will show up on the ECG. Some of these arrhythmas (or dysrhythmias) can be life-threatening because they greatly interfere with the heart's ability to pump blood around the body. Other arrhythmias, without being life-threatening, do require medical intervention such as drugs.

You may like to read about two arrhythmias, atrial fibrillation and ventricular fibrillation, and draw a diagram showing what the ECG pattern of each would look like. Your notes accompanying the diagrams should explain the myocardial activity of each arrhythmia and, in brief, the clinical signficance of each. Pay particular attention with atrial fibrillation to the altered shape of the P wave; and with ventricular fibrillation to the greatly altered shape and rhythm of the QRS complex. Which of these two arrhythmias do you think is the more dangerous, and why?

You may have noticed during your ward experience, or in some of the texts you consult, that an ECG is printed on a narrow strip of graph paper. The size of the squares on the graph paper is uniform in any clinical area, and so it is possible to calculate the heart rate by measuring the distance between two successive features such as the R waves (the long upward stroke of the QRS complex).

 ACTIVITY

Collect two examples of an ECG strip, showing perhaps five or six full cardiac cycles on each, one example showing tachycardia, the other normal rhythm. In which of your examples are successive QRS complexes closer together, and in which are they wider apart? However close together the QRS complexes appear, are they regular or irregular in rhythm?

Note: please liaise with your teacher for this activity, rather than invading your Coronary Care Unit or Intensive Care Unit with a large group of your fellow students. You may not be entirely welcome.

In order to record a patient's ECG, electrodes are placed at certain sites on the chest wall and the limbs. In this way, the voltage of the electrical activity of the heart can be 'looked at' from different directions by measuring it between two points. (The bigger the current, the larger the mark made on the graph paper.) To understand the complex nature of the ECG, you may find it helpful to begin by reading a comparatively basic description; you can then progress to more detailed texts. I make the following suggestions for reading, in order of increasing complexity: *Hubbard & Mechan 1987, pp. 118–119; Jennett 1989, pp. 157–161; Rutishauser 1994, pp. 85–86 and Fig. 5.11A–C; Hinchliff & Montague 1988, pp. 340–344; Guyton 1991, pp. 120–123.*

Make sure you read about both the limb leads and chest leads.

17. Draw a diagram representing the ECG leads in position on the body, and giving a typical ECG tracing for each of them. (One P–QRS–T series is sufficient for each lead.)

18. Why are the ECG tracings from different leads different in appearance? Compare, for example, lead aVR with lead I.

19. What does the term sinus rhythm mean?

SECTION 4: PRESSURE CHANGES AND HEART SOUNDS
Time for completion: about 2 hours

In an earlier Section we learned that heart valves open and close because of pressure changes within the heart chambers and the major blood vessels. You drew a series of diagrams showing blood flow through the heart that is somewhat similar to *Fig. 5.12* in *Rutishauser 1994*. But you'll also see in Rutishauser that at each stage of the diagram a graph shows how pressure alters in the left side of the heart.

20. Study this diagram step by step, and add brief notes to your own series of diagrams, in order to explain why certain valves open and close.

21. The next diagram in *Rutishauser 1994* (*Fig. 5.13*) provides more detail on the size of pressure changes, and also relates these changes to the stages of systole and diastole. Read what the text has to say about this figure, and then go on to read about the pressure changes in the right ventricle. Do you understand why the pressures on the right side of the heart are lower than those on the left?

FURTHER STUDY

There is more detail in *Hinchliff & Montague 1988, p. 346*. Read through this, and study *Fig. 4.2.28* only when you are sure you understand the description in Rutishauser. You'll see how several events – pressure changes on both left and right sides of the heart, valve movements, and the ECG – all relate to each other.

22. What are heart sounds? Doctors can listen to these by applying a stethoscope to the patient's chest. Note, in your reading, that these sounds are produced when the heart valves close, and that, in physiology texts, they are given the descriptive terms 'lubb' and 'dupp' (spelled variously). Note too that these sounds can change when there is damage to certain of the valves. See: *Mackenna & Callander 1990, p. 96; Rutishauser 1994, p. 88 and Fig. 5.13; Hinchliff & Montague 1988, p. 322, also Fig. 4.2.28.*

ACTIVITY

You have discovered what the term sinus rhythm mean ; but there is a type of heart beat called sinus arrhythmia. I should emphasise that this is, usually, quite harmless, and is normal in children and young adults.

Look at *Jennett 1989, p. 159, Fig. 6.7C*, for a rhythm strip showing sinus arrhythmia; and *Hinchliff & Montague 1988, p. 333, Table 4.2.1, and p. 344*. See also *Guyton 1991, p. 139*.

With a colleague, take each other's pulse and, as well as observing its rate (i.e. number of beats over a full minute) note its rhythm, too. Does your heart beat faster during inspiration, and more slowly during expiration? What is the explanation for this?

■ Blood vessels, pulse and blood pressure (guided study)

Time for completion   About 4 to 6 hours – up to 1 'college day'

Overall aim   To achieve an understanding of the different structures and functions of arteries, veins and capillaries, and of the nature of the arterial pulse and blood pressure.

Introduction   Taking a patient's pulse and blood pressure are perhaps the commonest observations made by nurses. Because they are done so often, it may be easy to forget their significance. Often, however, the physician or surgeon depends on accurate nursing observations to provide the information needed to make an important clinical decision.

By understanding the physiology of the pulse and blood pressure you will come to appreciate their importance in patient care, and why they are to be recorded for a particular patient.

Background questions
1. Examine a diagram showing the main arteries and veins of the body; see, for example, *Wilson 1990, pp. 82-83*. Find the aorta on this 'map' and follow it as it travels down through the thorax to the abdominal and pelvic cavities. Observe how it sends out branches to supply, for example, the kidneys, and the legs.

2. Similarly, on a map of the main veins of the body, follow the venous blood back to the heart from, for example, the intestines and liver, the kidneys, and the legs.

3. Making use of a simplified diagram such as *Wilson 1990, Fig. 5.19*; or *Rutishauser 1994, Fig. 5.16*, distinguish between the systemic and pulmonary circulations. Which of these is supplied by the aorta? Which flows into the venae cavae? Which returns blood to the left atrium?

Specific questions   **SECTION 1: ARTERIES AND ARTERIOLES, AND THE ARTERIAL PULSE**
Time for completion: 1 to 1½ hours

1. Arteries and veins share the same three types of tissue in their walls, though in different proportions. Draw a diagram showing a cross-section of both an artery and a vein, labelling the three tissue layers. See: *Wilson 1990, Fig. 5.1; Rutishauser 1994, Fig. 5.17*.

 COMMENTARY ON QUESTION 1

In each of the figures referred to above, you'll see that the vein has a somewhat flattened shape compared with the artery. Also, there is a difference in the amount of muscle and elastic tissue within the walls of the two types of blood vessel. Your diagram should show both these points.

---

2. Make notes on the changing structure of an artery, detailing its progression from major artery (e.g. the aorta) to arteriole. Note how the proportion of muscle and elastin in the middle layer of tissue changes.

3. Read about, and make notes on, the property of elastic recoil found in large arteries. At what stage of the cardiac cycle do these arteries distend the most, and at what stage do they recoil? What is the purpose of these changes? See: *Wilson 1990, Fig. 5.16; Rutishauser 1994, p. 94; Hubbard & Mechan 1987, p. 123.*

As arteries branch in their progress through the body, they become smaller in diameter and the proportion of elastic tissue becomes less. Conversely the proportion of smooth muscle in the artery walls increases. (Some texts distinguish between large elastic arteries and smaller muscular arteries.) The presence of this muscle enables the lumen (the internal cavity) of the artery to be increased and decreased – vasodilation and vasoconstriction.

4. Make brief notes on the nervous control of this smooth muscle, and how vasodilation and vasoconstriction are achieved. In this context, what is the meaning of the word tone?

5. Progressing further along the circulatory 'tree', arteries give way to arterioles. Draw a diagram of an arteriole in cross-section and compare it with your earlier cross-section of an artery. What tissue makes up the mass of the arteriole wall? How is this tissue innervated?

6. Vasoconstriction and vasodilation have a great effect on blood flow through the arterioles and, therefore, on the amount of blood reaching the capillaries. Which state – constriction or dilation – increases the resistance to blood flow? See *Rutishauser 1994, Fig. 5.21*, also the opening section of *Ch. 5, and Fig. 5.2.*

 COMMENTARY ON QUESTION 6

Note that it is the arterioles (rather than the arteries) that provide the highest resistance to blood flow. It is the arterioles rather than the arteries that influence the blood flow to the capillaries where the all-important exchange of nutrients and wastes between blood and tissues occurs.

---

7. You have already made brief notes on the nervous control of arterioles. Perhaps using a diagram such as Figure 3.12 make further notes on the nervous, hormonal and local controls of vasoconstriction and vasodilation. See: *Rutishauser 1994, p. 96; Hinchliff & Montague 1988, pp. 362–365; Mackenna & Callander 1990, p. 107; Guyton 1984, pp. 281–284.*

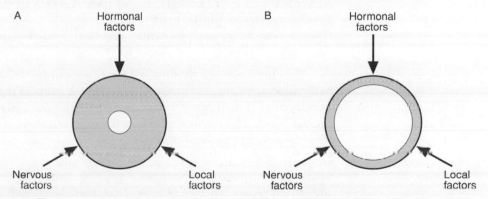

Fig. 3.12 Factors causing (A) vasoconstriction and (B) vasodilation. Write down as many as you can in each category.

FURTHER STUDY

Arteries and veins aren't just tubes conveying blood from and to the heart, though that is certainly part of their function. Arteries and veins have different structures and different properties. Veins – as we'll see in a later Section – are more distensible than arteries, and so they can act as reservoirs of blood. Arteries aren't passive carriers of blood from the heart, but control the blood supply to the tissues by widening or narrowing under the influences (nervous, hormonal and local) you've just described in your notes.

If you wish, read the opening section of *Guyton 1991, Ch. 14*, which provides fascinating insights into different parts of the circulation. Compare, for example, the percentage of blood contained by the arteries and arterioles with that in the veins and venules (*Fig. 14.1*). Guyton is a highly detailed text, but it is (usually) very readable. Don't be put off by the size and weight of the book.

When we talk of 'taking a patient's pulse' we mean his arterial pulse. In college you will learn about the various sites for recording the pulse and the significance of your observations.

8. What actually is the pulse? Is it the flood of blood rushing along the arteries after each ventricular contraction that you can feel? This sounds likely but it isn't the case. Also, what observations should a nurse make when recording a patient's pulse – that is, as well as the rate of the pulse? See: *Wilson 1990, p. 80; Hubbard & Mechan 1987, p. 127; Rutishauser 1994, p. 94; Jennett 1989, pp. 133–134; Hinchliff & Montague 1988, p. 367; Guyton 1991, pp. 160–163.*

9. Blood flow into the aorta and the pulmonary arteries occurs in surges, as blood is pumped out of the heart with each contraction. But by the time blood reaches the capillaries (having travelled through the circulatory 'tree' of smaller arteries and arterioles) these surges are usually absent, and the blood flow is smooth and continuous. Ensure that you understand how this is achieved by checking your notes on the ability of arteries to distend and recoil.

 This is an appropriate place to have a break.

SECTION 2: CAPILLARIES, VEINS AND VENULES
Time for completion: 1 to 1½ hours

(Note: you may wish to study 'Section 3: Blood pressure' before this part of the Work Sheet. It is possible to transpose Sections 2 and 3.)

At the farthest 'branches' of the vascular 'tree', and lying between the arterioles and the venules, is a massive network of tiny blood vessels called capillaries. It is from the capillaries that nutrients pass into the surrounding tissue fluid, and thence into tissue cells; while waste products from those cells pass in the opposite direction.

10. In your notes draw a diagram representing a capillary network. Ensure that your diagram shows both the arterial and venous ends of the network, and include arrows showing the direction of blood flow. See *Rutishauser 1994, Fig. 5.26.*

11. Make notes also on the structure of a capillary, especially the arrangement of endothelial cells and basement membrane. Show how capillaries can vary in different parts of the body, for example the brain, spleen, kidneys and skeletal muscle. See: *Rutishauser 1994, pp. 99–100; Guyton 1984, pp. 339–341.*

12. How is it that substances like nutrients and wastes can pass from capillary to tissue cells and vice versa? What is it that draws, for example, glucose from the capillary into the tissue fluid surrounding a cell, and then into the cell itself? What is it that draws waste products from a cell into its surrounding tissue fluid, and then into the capillary?

 COMMENTARY ON QUESTION 12

Take this question in easy stages, perhaps by using a series of simple diagrams representing a short length of capillary as in Figure 3.13.

First demonstrate hydrostatic pressure within the capillary, and how, in health, the pressure at the arterial end of the capillary is higher than at the venous end. Make notes on what factors create this hydrostatic pressure, what the normal pressure is, and how it pushes fluid out of the capillary into the surrounding tissue fluid. (Note how in most texts the unit used for the measurement of pressure is millimetres of mercury (mmHg). This is used in texts such as Guyton (1991). The SI unit, the kilopascal (kPa) is given as well as mmHg in Hinchliff & Montague (1988) so you may like to use both units in your own diagrams.)

Next, show how this hydrostatic pressure is opposed by osmotic pressure (sometimes called oncotic pressure) and how this is created. Give normal measurements for this osmotic pressure. You may like to base your notes on a simple diagram like Figure 3.14. See *Rutishauser 1994, pp. 100–102 and Fig. 5.27A & B.*

You may then like to show how substances such as glucose move from the tissue fluid into the cell, but you will need to read about how different substances cross cell membranes (*Rutishauser 1994, Ch. 2*).

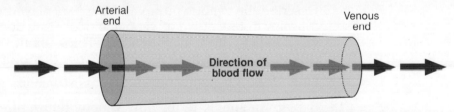

Fig. 3.13 Capillary showing arterial and venous ends.

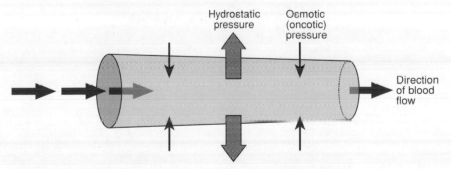

Fig. 3.14 Capillary showing hydrostatic and osmotic (oncotic) pressures.

13. With your knowledge of hydrostatic and osmotic pressures within the capillary, show how tissue fluid is formed. Make notes on what substances can, in health, leave the capillary for the tissue fluid, and what substances do not normally leave the capillary.

14. In a further diagram show the structure of lymphatic capillaries. What is their function regarding tissue fluid and the substances it contains? What happens to the fluid collected in the lymphatic capillaries? See: *Rutishauser 1994, Fig. 5.29; Mackenna & Callander 1990, p. 123; Hinchliff & Montague 1988, pp. 377–378; Guyton 1984, pp. 344–346; Guyton 1991, pp. 172–174.*

 FURTHER STUDY

Your notes so far have shown the pressures exerted on substances in the capillary in health, and how tissue fluid is formed. In some circumstances, however, changes occur in hydrostatic pressure at the arterial or the venous end of the capillary network, or osmotic pressure within the capillary or tissue fluid. Such factors lead to changes in the amount of tissue fluid formed. Look at: *Rutishauser 1994, p. 101 and Fig. 5.28; Hinchliff & Montague 1988, p. 379; Guyton 1984, pp. 347–349.*

(If using an American text such as Guyton, note that the American spelling of oedema is edema.)

After supplying the tissue fluid and cells with nutrients, and picking up waste products, blood flows from the capillaries into small venules and then into larger veins.

15. Show in your notes the structure of these vessels, demonstrating where the valves are to be found.

16. How does blood flow along the veins towards the heart, even when a person is standing up? (Bear in mind that in an adult the venous blood may have to 'climb' some considerable distance from his feet to the right side of his heart.) Describe the venous pump and the part played by the valves and the skeletal muscles in the return of venous blood. How does the structure of the walls of veins assist with venous return? (Think about the cross-section of a vein you drew earlier, and how it differed from that of an artery.)

17. How does a person's respiration assist venous return?

18. Explain in simple terms how venous return affects cardiac output.

 Take another break here.

SECTION 3: BLOOD PRESSURE
Time for completion: 1½ to 2 hours

All fluids held within a container exert pressure on the container's walls. Thus water in a swimming pool pushes against the walls of the pool. This can be experienced by a swimmer who, diving to the bottom, can feel the increase in pressure against the eardrums.

Blood, just like any other fluid, likewise exerts pressure against the walls of its container – in other words the blood vessels and the chambers of the heart. Physiologists distinguish between venous blood pressure, arterial blood pressure, and capillary blood pressure, but the most usual clinical observation made on the ordinary hospital ward, and in the patient's home, is the measurement of arterial blood pressure.

In this final Section of the Guided Study we'll concentrate on arterial blood pressure of the systemic circulation (rather than the pulmonary circulation). Bear in mind, though, that blood exerts pressure in vessels other than the systemic arteries.

19. First read 'Principles' at the beginning of *Rutishauser 1994, Ch. 5*, so that you're sure of the meaning of words like pressure and resistance. Then, as a brief background to blood pressure, turn to *Wilson 1990, pp. 79-80*.

20. Now read 'Arterial blood pressure' in *Rutishauser 1994, p. 97*. You'll note there the equation:

Blood pressure = Cardiac output × Peripheral resistance

By discovering what factors influence both cardiac output and peripheral resistance, we can gain a greater understanding of the complex subject of blood pressure. Have a look at Figure 3.15, which is my attempt at expressing some of the factors that maintain blood pressure. Copy this diagram, leaving space to add notes to each of the factors shown there. By all means adjust the diagram if you think it will make better sense drawn another way.

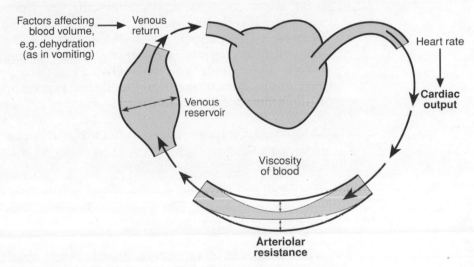

Fig. 3.15 Factors contributing to maintenance of blood pressure.

21. Swift changes to blood pressure are usually made via the nervous system acting on the arterioles (in other words, changing the peripheral resistance). We know from experience that our blood pressure is able to remain reasonably constant whether we are lying or standing, resting or exercising. In order for arterial blood pressure to be maintained within normal limits, it has to be monitored by the body. First, then, make notes on the arterial baroreceptors, where they're found, and how they work. How are they connected to the brain, and to which part of the brain? Look at *Rutishauser 1994, p. 98*, 'Control of blood pressure', and *Fig. 5.25*. See also *Guyton 1984, p. 310*, 'The baroreceptor pressure control system' (which also explains what happens when a person stands).

22. Now describe how nerve messages from the brain centres you've just studied

pass to the heart itself, and to blood vessels. What effect does sympathetic input have on the lumen of arterioles, and on the heart contractions? (See *Guyton 1984, pp. 308–310.*) For a neat summary of the control of arterial pressure, read the 'Overview' of *Ch. 19* in *Guyton 1984*.

23. How else are changes to peripheral resistance brought about? Read about the renin–angiotensin system, and the effect of angiotensin II on blood vessels, and add notes to your diagram.

24. Similarly make notes on the actions of adrenaline and noradrenaline, both on blood vessels and on the action of the heart itself. See *Hubbard & Mechan 1987, pp. 134–135,* for a very brief overview. More complex is *Mackenna & Callander 1990, pp. 108–109.*

    (You may find it useful to re-read the introduction to *Ch. 19* of *Guyton 1984,* where arterial pressure control is described briefly and clearly under three headings – Nervous control, Control by the kidneys, and Hormonal control.)

25. What is blood viscosity, and what part does it play in the maintenance of blood pressure?

 Have a short break here.

26. So far, we've concentrated on factors affecting the peripheral resistance. Changes in a person's cardiac output, however, can also affect his blood pressure. What is cardiac output? Partly it depends on the heart rate – the number of times the heart beats per minute. (What is the difference between stroke volume and minute volume?) Then there is a person's blood volume – the heart can only pump out the blood that is returned to it from the venous side of the circulation. See, for example, *Guyton 1984, pp. 284–285* and *Jennett 1989, pp. 138–140.*

    How might haemorrhage affect blood pressure? How does the body attempt to react to such an event in order to maintain blood pressure within safe limits?

27. You discovered earlier in this Guided Study how the veins can act as a reservoir for blood. If there were widespread vasodilation, what effect would this have on the cardiac output?

    Swift changes in blood pressure are brought about by nervous system control. Longer term adjustments involve the kidneys; this happens, for example, in chronic blood loss, as in a slow bleed from a duodenal ulcer.

28. How do the kidneys help in bringing about changes in blood pressure? See: *Rutishauser 1994, Ch. 13; Guyton 1984, pp. 312–313* (these pages cover the renin–angiotensin system for controlling arteriole size).

29. Make notes, too, on the action of antidiuretic hormone (ADH) and its effect on blood volume (and therefore on cardiac output). What stimulates the production of ADH, and what causes its production to be decreased? See: *Rutishauser 1994, Ch. 13; Mackenna & Callander 1990, p. 186.*

COMMENTARY ON SECTION 3

The maintenance of blood pressure is extremely complex, hence my suggestion for a diagram such as Figure 3.15 which attempts to summarise the various controlling factors. Note how many of the body's systems are active in this process – the nervous system, endocrine system, urinary and respiratory systems as well as the cardiovascular system.

Finally we consider how a person's blood pressure is measured clinically.

30. State the normal systolic and diastolic values of a young adult's blood pressure, and ensure you know what the terms systolic and diastolic mean in relation to blood pressure.

31. Descriptions of the recording of blood pressure using a mercury manometer are to be found in: *Rutishauser 1994, p. 97, and Fig. 5.23A & B; Hinchliff & Montague 1988, pp. 369–372.*

 ACTIVITY

In your reading about blood pressure, you'll discover that there are two levels on the mercury scale of the sphygmomanometer that could be taken to represent the diastolic pressure. These are: when the tapping sound (first heard at the systolic level) becomes muffled; and when it disappears altogether. With the help of members of your student group practice taking each other's blood pressure and noting all the sounds that occur. It's most important when on the wards to ensure you know which sound is taken to represent the diastolic blood pressure. Any change in a patient's blood pressure can then be related to his clinical condition rather than to lack of uniformity in recording his blood pressure.

 FURTHER STUDY

Related subjects which may usefully be studied include:

- pulmonary circulation pressures
- physiological changes in exercise.

For the latter, have a look at *Bursztyn 1990, Ch. 6,* 'Part 2 – The cardiovascular system in exercise', where there are some useful charts and diagrams. There is a later section on the effects of training on the cardiovascular system.

■ Heart and circulation (check quiz)

1. Explain how the sympathetic and parasympathetic nerves affect the rate and force of the heart's contractions. How do these two nerve supplies balance each other in health?

2. A friend of yours has heard of artificial pacemakers but doesn't understand the function of the heart's own pacemaker, the sino-atrial node. Explain how this node functions, and why it is referred to as the pacemaker of the heart.

3. A patient has been admitted to your ward following a prolonged nosebleed (epistaxis). She complains of feeling cold and faint. How do you think these symptoms might arise, and what changes would you expect in her blood pressure and pulse? Why have these changes occurred?

4. You are showing a new care assistant how to take a patient's blood pressure, and she asks you why this observation consists of two numbers. How would you explain the systolic and diastolic pressures to your colleague, and what would you stress to her about the diastolic reading?

5. What adaptations in blood pressure does the body make when a person gets up after lying down for some time? What happens to the blood pressure when a person faints?

6. Describe the differences to be observed between arterial, venous and capillary haemorrhage. (Capillary haemorrhage may be seen if a child falls and superficially scrapes his knee.)

    A patient on your ward has had a specimen of arterial blood taken for analysis of blood gases. How would you care for the puncture site after the needle is withdrawn? You may like to discuss this with your ward mentor.

7. A fellow student in your set finds it difficult to envisage how the veins can act as a reservoir for blood. How would you explain to him this particular function of the veins? Give one condition or situation where blood can be shifted out of this venous reservoir or pool.

8. You notice that one of your patients, an elderly lady with heart failure, has swollen ankles after she has been sitting in a chair watching television for several hours. Your mentor explains that this is oedema, and the patient tells you her ankles are usually swollen at the end of the day. Why do you think this clinical feature might occur? (Don't worry too much about the medical condition of heart failure, but think about how tissue fluid is formed, and the various pressures – hydrostatic and oncotic – exerted in the capillaries.)

■ Respiration — the mechanics of breathing (work sheet)

Time for completion
About 3 to 5 hours

Overall aim
To explore how the structure of the respiratory system enables air to be drawn into and expelled from the lungs.

Introduction
The purpose of respiration is to ensure that the living cells of the body are provided with an adequate supply of oxygen, and that carbon dioxide, a waste product of their metabolism, can be excreted. A later Guided Study examines how gases are exchanged between lung and blood, deep inside the tissues of the lung. Here we look at how air is breathed in and out, and how the structure of the respiratory system enables breathing to occur efficiently.

Breathing is one of our most vital activities. If breathing ceases, unconsciousness soon intervenes, and cells begin to die within a very few minutes. People become naturally very anxious if they experience difficulty with breathing (as in asthma). The nurse can help by positioning the patient so as to facilitate breathing, and by administering and monitoring oxygen and prescribed drugs. Knowledge of the physiology of respiration is essential for an understanding of how correct patient positioning helps breathing, and how the drugs that are prescribed work.

Background questions
1. Describe the structure of the nose and nasopharynx, and explain what effect they have on inhaled air.

2. Describe the structure of the larynx. What is its role during:

   - speaking
   - swallowing?

3. Many children and young adults suffer from tonsillitis. Where are the tonsils found and what is their function in childhood?

4. Describe the structure of the trachea, and its relationship to the larynx, the lungs, and the oesophagus.

   For Questions 1 to 4 see *Wilson 1990, Ch. 7.*

Specific questions
GETTING AIR INTO THE LUNGS
Time for completion: 1 to 2 hours

1. What is the composition of atmospheric air? Find out the percentage of:

   - oxygen
   - carbon dioxide.

   Which inert gas makes up the greatest part of the atmosphere? Add it to the other two gases to get the composition of the air that we breathe. Later you'll be noting the composition of the air breathed out. See *Rutishauser 1994, p. 148.*

2. Check your notes on the basic gross structure of the respiratory tract from nose to lung, and draw a diagram showing the various parts. Note how the right and left main bronchi differ slightly in appearance. You should include bronchioles, terminal bronchioles and alveoli in your diagram. Have a look at *Marieb 1992, Fig. 23.6(b)*, which shows a cast of the bronchial 'tree', and demonstrates the delicate structures deep inside the lungs.

3. Draw a new diagram illustrating the microscopic structure of the alveoli, showing the cells making up their walls, the glands that secrete mucus, and the cilia. Add brief notes on the function of the cilia.

4. Draw a further diagram illustrating the pleura (both visceral and parietal layers). Read about the pleural cavity which exists between the two layers, and add notes on the importance of the pleura and the pleural fluid that is secreted into the pleural space. See *Wilson 1990, Fig. 7.17*, for a clear demonstration of the twin layers of pleura.

5. Now read about, and make notes on, the ribs and intercostal muscles, and the diaphragm, and their functions during respiration. See *Rutishauser 1994, Figs 7.4 and 7.5*.

    Notice how the dimensions of the chest are increased during inspiration (breathing in) and note what happens to both the diaphragm and the ribs, and to the pleura. As the chest cavity increases in size during inspiration, what happens to the pressure inside the lungs in relation to atmospheric pressure?

6. Note the role of the accessory muscles of respiration.

COMMENTARY ON QUESTION 6

When nurses position patients who are having difficulty in breathing, they are attempting to help them utilise these accesory muscles. For example, some patients like to sit well up in the bed with their back well supported and their arms resting on a bed table in front of them. Try to work out how this position would help the accessory muscles to be used.

7. Usually we aren't conscious of any effort in breathing, unless we've been running hard or have a chest infection. Yet, in health, inspiration is said to be an active process, while expiration is a passive process. Make sure you understand what these terms mean. You'll need to discover at what stage of respiration the various muscles contract, and what is meant by the recoil of the lungs. See: *Rutishauser 1994, p. 140; Wilson 1990, pp. 133–134, with Fig. 7.23*.

Later on in this Work Sheet we'll look at the nervous control of breathing. Before concluding this part of the Work Sheet, though, look back at your notes and diagrams, which together should give you a clear summary of how air is drawn into the lungs during inspiration, and expelled during expiration.

LUNG VOLUMES
Time for completion: about 1 hour

When we're breathing normally, the chest rises and falls gently. When we exercise, respirations become deeper and so the chest movements are more pronounced. In some diseases affecting the chest, the physician needs to know how deep a breath a patient can take, and how quickly and forcefully he can expel it. In this Work Sheet we're not concerned with disease processes, but in this part we'll examine the different measures of lung volume that can be made.

Volumes of air breathed in and out can be measured on an instrument called a spirometer (see *Rutishauser 1994, Fig. 7.9*).

8. What is the normal volume of air breathed out during quiet, resting respirations, and what name is given to it? (The name actually suggests flow in and out.)

    A spirogram of this gentle flow of air might appear as shown in Figure 3.16. Note how the spirogram shows not just the volume of air breathed in and out, but the rate of respirations as well.

9. How can minute respiratory volume be calculated?

10. Suppose that a person whose respirations are being measured by spirometer takes a very deep breath in, followed by a very deep breath out. What names are given to these two volumes of air? Show them on a new diagram like Figure 3.17.

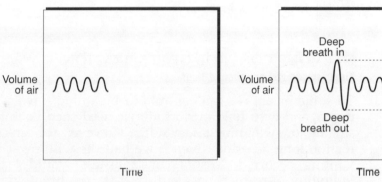

Fig. 3.16 Flow of air during gentle breathing.

Fig. 3.17 Flow of air during deep breathing.

The total volume of air moved in and out of the lungs during a deep inspiration and deep expiration is called the vital capacity (see Fig. 3.17). What is the normal vital capacity for a healthy adult? How do you think this might compare with the vital capacity of an athlete?

 FURTHER STUDY

There are some diseases that reduce a person's vital capacity, perhaps by weakening the respiratory muscles (as in poliomyelitis), by 'fixing' the spine and ribs so that their range of movement is limited (as in some forms of arthritis), or by making the lung tissue itself 'stiffer' or more difficult to inflate (as in pulmonary oedema). So a patient's vital capacity can be an important diagnostic measurement for the doctor.

You may like to read further about lung compliance and the role of surfactant in *Rutishauser 1994, Ch. 7. Hinchliff & Montague 1988, Ch. 5.3*, contains further information about lung volumes and the effects of certain disease processes.

11. At the end of even the deepest expiration, the lungs are not completely empty. What is the term given to the volume of air that is left? Find out the name given to the volume of air remaining in the lungs at the end of a quiet expiration. Add these two volumes to your spirograph and compare it to that in *Mackenna & Callander 1990, p. 136*.

12. Read about the recoil forces of the lungs, which act to push air out of the

lungs during expiration, in *Rutishauser 1994, p. 145*. Note the role of the pleura in expiration, and the changes in intrapleural pressure during the respiratory cycle.

13. Only part of the air moving in and out of the lungs is physiologically useful; that is, it participates in gas exchange. The great majority of gas exchange occurs deep in the lungs – in the alveoli. Air in the trachea and main bronchi, for example, plays very little part in gas exchange. For this reason, this non-functional part of the respiratory tract is called dead space. Find out the average volume of dead space, and show it on your diagram of the respiratory tract.

 Take a break here.

NERVOUS CONTROL OF RESPIRATION
Time for completion: about 1 hour

To some extent we can control our breathing – we can speed it up or slow it down, and even hold our breath; we need such voluntary control for speaking, singing, or swimming under water. However, we can't choose to stop breathing for too long a period. Even if we had the willpower to hold our breath until fainting occurred, loss of consciousness would lead to loss of voluntary inhibition of respirations, and we'd start to breath again.

14. Make a list of factors that may cause us to alter our rate and depth of breathing, either voluntarily or involuntarily. Exercise is one, and sleep another. How many more can you find?

15. Draw a diagram showing the nerve supply to the diaphragm and the intercostal muscles. Show at which level these nerves leave the spinal cord, and make notes on the implications of this for a person whose spine is fractured.

16. With regard to involuntary control of breathing, make notes on the respiratory centres in the medulla and pons.

For Questions 14 to 16 see *Rutishauser 1994, Figs 7.7 and 7.8* and *Wilson 1990, Fig. 7.25*. Compare your notes and diagram with the more complex diagrams in *Mackenna & Callander 1990, pp. 143 and 145*.

17. Distinguish between the pneumotaxic centre and the apneustic centre, stating briefly what function each has.

18. Describe the role of the inspiratory centre in the medulla. What is its relationship to the pneumotaxic centre?

19. Describe the role of the expiratory centre, also found in the medulla. Does the expiratory centre play a part in quiet expiration?

For Questions 17 to 19 see *Mackenna & Callander 1990, p. 143*.

20. Finally, describe the nerve receptors that are stimulated by:

- lung expansion
- the inhalation of an irritant like a biscuit crumb.

Where are these receptors found and, when stimulated, what effect do they have?

■ Respiration — gas exchange (guided study)

Time for completion   About 2 'college days', with 5 to 6 hours' study per day

Overall aim   To develop the student's background knowledge of the respiratory system, by examining how oxygen is conveyed from the alveoli in the lungs to the cells of the body, and how carbon dioxide is returned to the lungs for excretion.

Introduction   The previous Work Sheet looked at the mechanics of respiration – the muscular effort entailed in drawing air from the atmosphere into the deep tissues of the lung. It's here in the alveoli that the function of gas exchange occurs.

In this Guided Study we look at why gas exchange occurs, how the structure of the alveoli and their capillaries contributes to that vital function, and how gas levels in the blood control respiration rate and depth. We look at terms like diffusion and partial pressures of gases in order to understand how gases like oxygen and carbon dioxide behave within the lungs and blood as they do.

Background questions   1. Remind yourself of the composition of atmospheric air from the previous Work Sheet, in particular the percentages of oxygen and carbon dioxide. Note how the percentage of carbon dioxide is practically nil in atmospheric air. Read about the role of water vapour in atmospheric air: in *Table 7.2* of *Rutishauser 1994* it is assumed that no water vapour is present in the atmosphere.

2. Find out the composition of alveolar air and compare it to that of atmospheric air. Note the addition of water vapour to alveolar air. See how the amount of oxygen and carbon dioxide has changed between the atmosphere (the air that's breathed in) and the air deep within the lungs.

3. You'll now have your own table showing compositions of atmospheric and alveolar air, but so far it will show only the percentages of various gases. Read about the partial pressure of a gas: what the term means and how, with reference to oxygen and carbon dioxide, it is calculated. Add the partial pressures of oxygen and carbon dioxide to your table, for both atmospheric and alveolar air. As well as *Rutishauser 1994, Table 7.2*, see *Mackenna & Callander 1990, pp. 137 and 138*.

The latter text gives the composition of expired air as well as atmospheric and alveolar air. See if you can understand why the oxygen and carbon dioxide content of expired air lies something like midway between their levels in atmospheric and alveolar air.

Specific questions   ## SECTION 1: CROSSING THE ALVEOLAR–CAPILLARY BARRIER
Time for completion: 1 to 2 hours

First we need to discover what separates the oxygen and carbon dioxide molecules in the alveolar air and the gas molecules in the blood.

1. Draw a diagram representing the capillary supply to the alveoli. See, for example, *Wilson 1990, Fig. 7.20*. Show how the capillaries are wrapped very closely around the alveoli.

2. Now draw a diagram showing a cross-section of one pulmonary capillary and an alveolus. Note the epithelial cells making up the walls of both blood vessel

and alveolus, and their basement membranes. See: *Rutishauser 1994, Fig. 7.14; Jennett 1989, Fig. 8.15;* and *Guyton 1991, Fig. 39.9.*

As air moves in and out of the alveoli, and as blood flows through the pulmonary capillaries, oxygen diffuses from the air into the blood, while carbon dioxide diffuses from the blood into the alveoli (in order to be breathed out from the lungs).

3. Look at *Rutishauser 1994, Fig. 7.15,* which illustrates what is meant by the term diffusion, and read the section: 'Diffusion of gas molecules into liquid'.

4. Now move on to read 'Diffusion of oxygen and carbon dioxide' in Rutishauser, and see the accompanying figure (*Fig. 7.16*). Note how gas molecules move from a higher partial pressure to a lower, and that it is this pressure gradient (see *Fig. 7.17*) that brings about the transfer of gases in the lungs.

You may also like to cover the same ground in: *Marieb 1992, pp. 747–750; Jennett 1989, pp. 204–205;* and *Guyton 1984, pp. 446–447.*

5. Can you construct your own simplified diagram to illustrate this movement of gases from alveolus to capillary? *Rutishauser 1994, Fig. 7.16,* or *Wilson 1990, Fig. 7.24* may be helpful.

 FURTHER STUDY

How quickly does oxygen move from alveolus to capillary? How much time does the gas have to make this transfer if the person is exercising? Read *Guyton 1991, pp. 433–434,* 'Uptake of oxygen by the pulmonary blood'; a short and very accessible passage. You'll probably be surprised at how quickly oxygen can diffuse across the cell walls.

 Take a short break here.

SECTION 2: VENTILATION AND PERFUSION
Time for completion: 1 to 1½ hours

Oxygenation of body cells – which is the purpose of respiration – depends on getting air in and out of the lungs (ventilation) and on having a good circulation of blood through the lung tissues (perfusion). A good flow of air in and out of the lungs is of little use if the circulation of blood is poor, because oxygen, even though present in the alveoli, cannot be transported to the tissue cells.

In this Section of the Guided Study, we look at the balance between ventilation and perfusion. Once again, the main text to have with you is *Rutishauser 1994, Ch. 7.*

6. Remind yourself of the flow of blood from the right side of the heart via the pulmonary arteries; see *Rutishauser 1994, Fig. 7.18.* Note that, although the pulmonary circulation has a much lower pressure than the systemic circulation, the amount of blood flowing through the lungs is the same as the cardiac output – the amount of blood pumped out of the left ventricle. (Remember that the left ventricle can only pump out the amount of blood that it receives from the lungs and the right side of the heart.)

How many litres of blood per minute flow through the lungs, both in a resting person and one who is exercising?

7. Find out the pulmonary arterial pressure and compare it with the systemic arterial pressure. What is the average pressure within the pulmonary capillaries? (As well as Rutishauser, you might like to read 'Pulmonary Vascular Pressures' in *Guyton 1984, Ch. 27*.)

8. The pulmonary arteries differ in a number of ways from arteries elsewhere in the body. They are comparatively thin-walled, and so can act as reservoirs of blood, and their lumen alters in the close presence of oxygen. Find out whether pulmonary arteries dilate or constrict when oxygen levels nearby increase, what effect this has on pulmonary blood flow, and how systemic arteries (by contrast) react to higher oxygen levels. Both these points are covered in *Rutishauser 1994, Ch. 7*; see also *Guyton 1984, p. 448*.

9. With your knowledge of pulmonary blood vessels, now read about ventilation–perfusion balance in *Rutishauser 1994, p. 151*.

 COMMENTARY ON QUESTION 9

Your notes should include reference to physiological and anatomical dead space, and to shunts. *Rutishauser 1994, Fig. 7.19*, demonstrates both how a perfusion–ventilation mismatch occurs, and how physiological changes can bring about an improved balance.

 FURTHER STUDY

It may help your understanding of the interaction between perfusion and ventilation if you read about conditions where mismatch occurs – as in chronic obstructive airways disease (COAD). This is explained in *Rutishauser 1994, p. 152*, where you'll find not only a description of the disease processes but their relation to altered physiology. *Guyton 1984, pp. 449–451*, explains the effect of certain congenital heart defects on pulmonary circulation. (You may have heard of hole in the heart babies or blue babies – conditions which are described here.)

There is a very detailed section on ventilation and perfusion in *Jennett 1989, pp. 215 220*, which doesn't cover disease processes. Before tackling this text, however, be sure you fully understand the rather more accessible descriptions of Rutishauser and Guyton.

 Take another break here.

SECTION 3: OXYGEN TRANSPORT IN THE BLOOD
Time for completion: 2 to 3 hours. Aim to complete this Section at the end of your first 'college day' devoted to this Guided Study.

First let's review what we've covered so far in discovering events in the respiratory cycle. We have seen how air is drawn into the lungs by muscular effort and how it is breathed out, normally by the relaxation of the respiratory muscles. We've compared the contents of atmospheric air (the air that's breathed in) with the air found deep inside lung tissue (alveolar air). Oxygen has crossed the membranes of alveoli and capillaries to replenish de-oxygenated blood, and carbon dioxide has crossed in the opposite direction in order to be expelled. Remind yourself what it is that causes this movement of oxygen and carbon dioxide across the membranes of alveoli and capillary.

Now that oxygen has got into the capillary blood surrounding the alveoli, it has to be carried to cells throughout the body in order to replenish them. It is this carriage of oxygen in the blood that we'll investigate now, and then the carriage of carbon dioxide in the next Section.

10. By far the greatest amount of oxygen is carried attached to the haemoglobin of red blood cells, and this is what we'll be looking at in this Section. First, however, how is the remaining 1.5 to 2% of oxygen carried in arterial blood? In which part of the blood is this small amount of oxygen carried?

11. Now we turn to the vital role of haemoglobin, a complex structure with its protein chains and iron atoms. Draw a simplified diagram showing a haemoglobin molecule, and demonstrating the iron atoms to which oxygen becomes attached. Show on this diagram how much oxygen can be carried by each haemoglobin molecule. You could base your diagram on *Fig. 7.21* in *Rutishauser 1994*.

12. What is meant by a haemoglobin molecule being saturated with oxygen? What quantity of oxygen (in millilitres) can be carried by 1 gram of fully saturated haemoglobin? How much oxygen is carried in 100 ml of blood, providing that the blood has a normal haemoglobin level?

13. What term is given to the compound of haemoglobin and oxygen; and what term to haemoglobin that has given up its oxygen? What is the difference in colour between these two types of haemoglobin? What significance might this have for the colour of both arterial and venous blood?

 FURTHER STUDY

Oxygen combines very readily with haemoglobin but another gas, carbon monoxide (CO), combines even more readily. Read about how CO poisoning can dramatically lower a person's PO_2. What is the appearance of carboxyhaemoglobin and how does this affect the appearance of someone suffering from CO poisoning? What are two major sources of CO? *Rutishauser 1994, p. 154*, mentions several other factors that can adversely affect haemoglobin's ability to convey oxygen.

How much oxygen can be carried in a person's blood as it leaves his lungs? This will depend partly on the number of red cells he has, and the amount of haemoglobin within those red cells. (What name is given to the common medical condition whereby a person's haemoglobin level is low?) A major factor, though, in the carriage of oxygen is the partial pressure of oxygen (PO_2). It's the pressure gradient between alveolar air PO_2 and pulmonary capillary blood PO_2 that brings about the diffusion of oxygen from alveoli to blood, as you discovered earlier. So it is the PO_2 of the pulmonary blood that affects how much oxygen is taken up by the haemoglobin. This attachment –

and detachment – of haemoglobin and oxygen is usually expressed in the form of a graph: see *Rutishauser 1994, Fig. 7.22.* The curve, called the oxygen–haemoglobin dissociation curve, is said to be shaped rather like an S (though like me you may find it rather difficult to see the resemblance). This curve demonstrates how much oxygen is taken up by haemoglobin when the PO_2 is high, and when the PO_2 is low. Figure 3.18 shows a very simplified version of the curve.

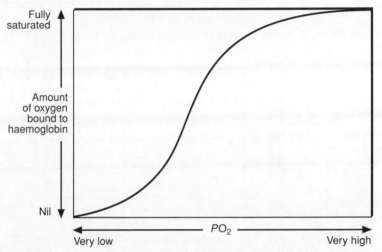

Fig. 3.18 The amount of oxygen carried by haemoglobin (the S-shape is exaggerated).

I haven't attached any figures to this graph; what we'll do now is build up our own complete graph step by step.

14. First construct the 'frame' for our graph, the vertical axis showing the percentage of oxygen attached to haemoglobin, the horizontal axis showing the partial pressure of oxygen (PO_2) (see Fig. 3.19). Note that whereas British texts like Rutishauser use kilopascals (kPa) as units of partial pressure, American texts, including Guyton and Marieb, use millimetres of mercury (mmHg).

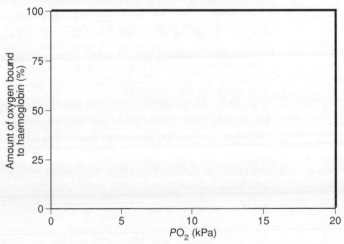

Fig. 3.19 First stage of oxygen–haemoglobin dissociation graph, showing PO_2 and haemoglobin saturation.

15. In the pulmonary capillaries, PO_2 is high. The PO_2 of blood just leaving the lungs is usually put at 13.2 kPa. With a partial pressure as high as this,

haemoglobin normally achieves a high saturation level – about 97%. This means that most of the binding sites in the haemoglobin molecules are occupied by oxygen. Add a point to your graph, showing haemoglobin saturation as 97% and PO_2 as 13.2 kPa (see Fig. 3.20).

16. Let's now move to some tissue where the PO_2 is low – perhaps 3 kPa. At this level, the oxygen–haemoglobin dissociation curve (as in *Rutishauser 1994, Fig. 7.22*) shows us that the haemoglobin readily gives up its attached oxygen, in other words, the oxygen dissociates from the haemoglobin. The haemoglobin saturation level, with a tissue PO_2 of only 3 kPa, may well be only 45%. Show this on your own graph, as in Figure 3.21.

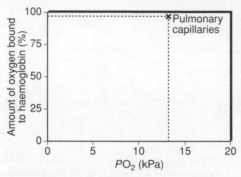

Fig. 3.20 Saturated haemoglobin when PO_2 is high. Hb saturation = 97%; PO_2 = 13.2 kPa.

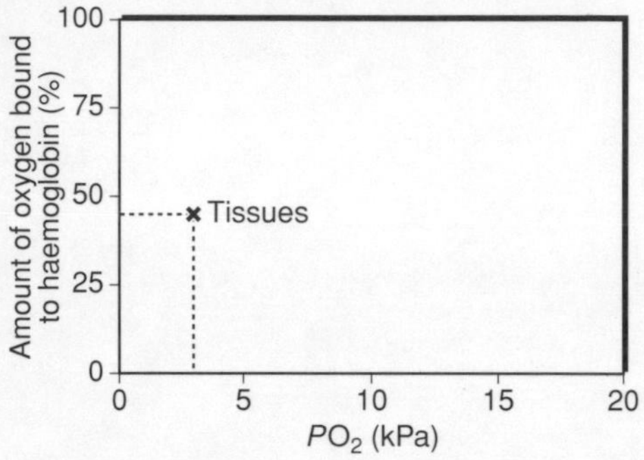

Fig. 3.21 Low saturation level of haemoglobin when PO_2 is low. Hb saturation = 45%; PO_2 = 3 kPa.

An obvious question now is, why can't we simply draw a straight line on our graph connecting the two points. Such a line would show, incorrectly, that the rate at which haemoglobin gives off its attached oxygen is regular. In fact, the rate at which haemoglobin releases oxygen is both irregular (hence the S-shaped curve) and beneficial to the human body.

17. As blood leaves the lungs – well oxygenated, remember – the PO_2 of the surrounding tissues becomes lower. The PO_2 of venous blood is normally about 5.3 kPa, but at this level haemoglobin is still about 70% saturated (see Fig. 3.22). Add this new point to your graph.

It's quite surprising to discover that so-called de-oxygenated blood still has a haemoglobin saturation level of 70%.

18. But tissue PO_2 can fall lower than 5.3 kPa; for example in heavy exercise the PO_2 of some tissues can fall to about 1.5 to 2 kPa. When this occurs, haemoglobin molecules in the blood supplying those tissues gives up their oxygen far more readily. This explains why the dissociation curve becomes steeper as the oxygen partial pressure falls especially low. Show this by adding a new reading to your graph, where PO_2 is 1.5 kPa, and the haemoglobin saturation level is approximately 17.5% (see Fig. 3.23).

19. At the upper end of the curve, where PO_2 has fallen to 10 kPa, the haemoglobin saturation is still as high as about 95%. Add this point to your graph. Now read the explanation given in Rutishauser about further increases (i.e. above 13.5 kPa) of oxygen pressures. Haemoglobin saturation cannot increase beyond 100%, which explains the long flattened line adopted by the graph in *Rutishauser 1994, Fig. 7.22*. Add this to your own graph and you'll see that you have yourself produced something approaching the S-shaped oxygen–haemoglobin dissociation curve (see Fig. 3.24).

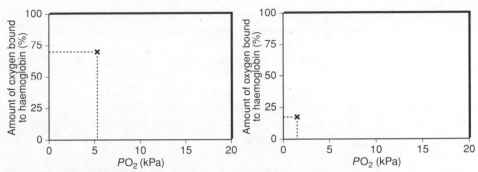

Fig. 3.22 Haemoglobin saturation when PO_2 is 5.3 kPa (as in venous blood leaving the tissues). Hb saturation = 70%.

Fig. 3.23 Haemoglobin saturation level for tissues with a PO_2 of only 1.5 kPa. Hb saturation = 17.5%.

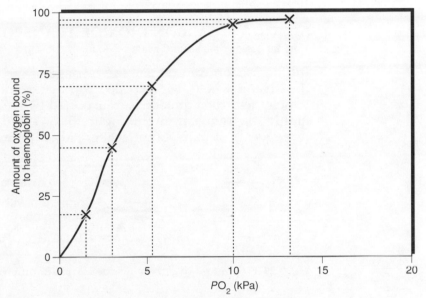

Fig. 3.24 The oxygen–haemoglobin dissociation curve complete.

This would be a good place to review your notes and the graph you've completed. You should now be clear how the S-shape of the dissociation curve expresses the rate at which haemoglobin gives off its oxygen.

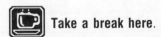 **Take a break here**.

20. Although the carriage of carbon dioxide is covered in the next Section of this Guided Study, the effects of carbon dioxide levels on oxygen–haemoglobin dissociation are relevant here. Higher levels of carbon dioxide enable haemoglobin to off-load its oxygen more readily. This means that when oxygenated blood reaches the tissues where carbon dioxide levels are comparatively high, this higher PCO_2 enhances the haemoglobin's release of oxygen. Read about this in *Rutishauser 1994, p. 154*, and note how this factor of oxygen dissociation is shown in *Fig. 7.22*. The somewhat complex diagram in *Mackenna & Callander 1990, p. 139*, also illustrates this, and gives details of other factors besides CO_2 that enhance the dissociation of oxygen. (The same ground is covered in Hubbard & Mechan 1987 and Jennett 1989; see the references given in the suggestions for further study that follow.)

FURTHER STUDY

The oxygen–haemoglobin dissociation curve is such an important part of the physiology of respiration that it is well worth finding out what other writers have to say on the subject. Try: *Hubbard & Mechan 1987, pp. 157–159; Jennett 1989, pp. 206–210* (a somewhat technical description); *Guyton 1984, pp. 451–454* (my own preference for clarity, though be warned that partial pressures are here expressed only in millimetres of mercury (mmHg)).

This is the place to end your first 'college day' with this Guided Study.

SECTION 4: CARBON DIOXIDE TRANSPORT IN THE BLOOD
Time for completion: about 3 hours

In this Section we examine the carriage of carbon dioxide from tissue cells to the alveoli where it is expelled during expiration. There are three forms in which the carbon dioxide is transported by the blood, and these are expressed highly diagrammatically in Figure 3.25. The diagram shows that each of the three forms of transport affects greatly differing amounts of carbon dioxide.

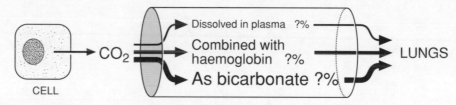

Fig. 3.25 Proportions of carbon dioxide carried in different forms in the blood from cells to lungs.

21. The smallest percentage of carbon dioxide is carried directly dissolved in the plasma. Add this percentage to the diagram.

22. A larger percentage of carbon dioxide is carried combined with haemoglobin and with plasma proteins. Add this percentage to the diagram.

23. By far the greatest proportion of carbon dioxide carried in the blood is in the form of bicarbonate. Add this percentage to the diagram.

24. The carbon dioxide is combined very loosely with haemoglobin. What is the significance of this loose bond when the carbon dioxide reaches the capillaries surrounding the alveoli?

The remainder of this Section is concerned with the carriage of carbon dioxide as bicarbonate. This is a complex subject so, as with previous complicated topics, it will be covered in gradual stages.

25. Begin a series of small-scale diagrams by copying Figure 3.26A. This shows carbon dioxide entering a red blood cell. Inside the red cell the CO_2 combines with water (H_2O) to form H_2CO_3. Add the name of this acid to your notes attached to this first small diagram, and also add the name of the enzyme inside the red cell that greatly hastens this reaction by about 5000 times.

26. Now the H_2CO_3 dissociates into two ions: hydrogen (H^+) and bicarbonate

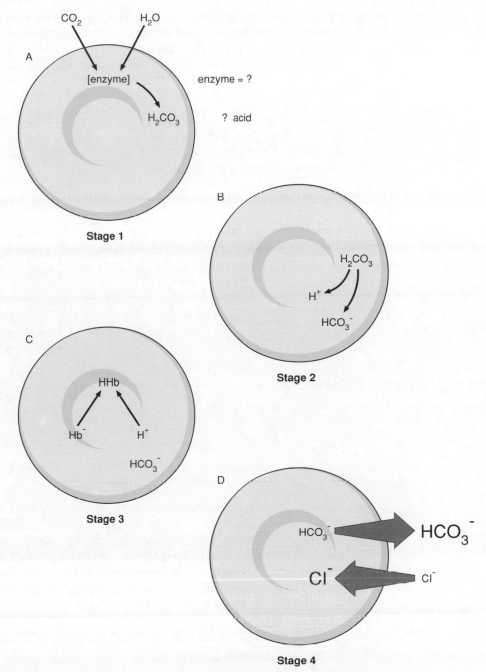

Fig. 3.26 (A) CO_2 enters the red cell, and joins with H_2O to form an acid. (B) The acid dissociates into hydrogen and bicarbonate ions. (C) Most of the hydrogen ions combine with haemoglobin (Hb). This leaves the bicarbonate ions free to (D) move out of the cell while chloride ions move in.

(HCO_3^-). One of these is an anion, and the other a cation. Check on the meaning of these terms, and copy Figure 3.26B into your notes.

27. The hydrogen ions then combine with haemoglobin within the red cell. In this way haemoglobin, which is a powerful acid–base buffer, 'mops up' excess hydrogen ions which can cause a solution to be acid. (If you want to read about how hydrogen ions make a solution acid, see *Rutishauser 1994, Ch. 12.*) Show this third stage of the process in your next small diagram, as in Figure 3.26C.

28. Much of the bicarbonate part of H_2CO_3 now diffuses into the plasma. Because this outward diffusion of a negatively charged ion, HCO_3^- (anion or cation?), would cause an electrical inbalance, chloride ions, Cl^-, diffuse into the red cell. Show this dual movement in your next diagram as in Figure 3.26D (p. 131).

    What is it that helps this outward movement of HCO_3^- and inward movement of Cl^- to occur so quickly? See *Rutishauser 1994, Fig. 7.23*.

29. This movement of chloride inside the red blood cell is termed the chloride shift. Now put together all your small diagrams to create a large-scale diagram illustrating the chloride shift. Figure 3.27 starts you off. Complete your diagram by filling in the boxes.

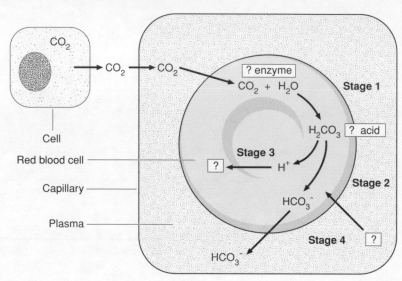

Fig. 3.27 The chloride shift.

30. Look at *Rutishauser 1994, Fig. 7.24*, and read about the carbon dioxide dissociation curve. How does this differ from the oxygen dissociation curve studied earlier? Copy out the diagram in Rutishauser (or any other physiology text you are using) and add to it the normal range of PCO_2 levels (as in *Rutishauser 1994, Fig. 7.16*) from that of the blood leaving the lungs (relatively low PCO_2) and that of blood arriving at the lungs (relatively high PCO_2). Note how narrow the band of normal PCO_2 values is.

 FURTHER STUDY

There is a relationship between the carriage of oxygen and carbon dioxide via the red blood cells. If CO_2 is high (as in the tissues) this causes O_2 to be released from haemoglobin, thus 'making room' for CO_2 to be taken on. When O_2 is high (as in the lungs) CO_2 is released from haemoglobin, thus allowing O_2 to be taken up by the haemoglobin. This is expressed very simply in Figure 3.28. Read about this relationship in *Guyton 1991, pp. 441–442*, the Haldane effect.

It is well worth reading about carbon dioxide transport in a number of different texts. The following references may be useful: *Hubbard & Mechan 1987, pp. 160–161; Jennett 1989, pp. 212–213* (you could also read *pp. 213–214* on over- and under-breathing); *Hinchliff & Montague 1988, pp. 496–497* (for interest, see also *pp. 497–498* on hypocapnia and hypercapnia – check your understanding of these terms).

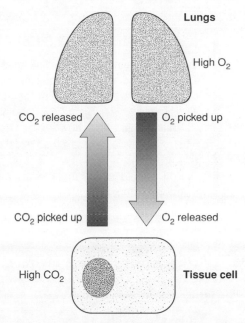

Fig. 3.28 Relationship between oxygen transport and carbon dioxide transport.

 This is a suitable place for a well-deserved break.

SECTION 5: GETTING OXYGEN TO THE TISSUES
Time for completion: 2 to 3 hours. This Section should complete your second 'college day' on this Guided Study.

In this Section we learn about the chemical control of respiration. You will already appreciate that factors such as the cardiac output, and the red cell count and haemoglobin level, will have an effect on the oxygenation of cells. (If there are fewer red cells in my blood, there is less oxygen-carrying capacity. Similarly, if my pulse is especially slow, less oxygen will be delivered to the tissues than if the pulse is normal.) These factors, however, will be left for your further study.

Instead, this Section is concerned with the controlling effect of oxygen, carbon dioxide, and hydrogen ion concentration on respiration.

31. Referring to *Fig. 11.3A* in *Rutishauser 1994*, draw a diagram showing the position of chemoreceptors, and their nerve connections to the pons and medulla. (You could also refer to *Wilson 1990, Fig. 7.25*; and *Guyton 1991, Fig. 41.4*.)

32. Read about, and make notes on, the effect of PCO_2 on these chemoreceptors. See *Rutishauser 1994, pp. 214 and 229*.

33. Similarly, read about, and make notes on, the effect of changes in pH and PO_2 on these chemoreceptors.

See page 134 for a Commentary on Questions 32 and 33.

COMMENTARY ON QUESTIONS 32 AND 33

Your notes should clearly distinguish between central chemoreceptors (which Guyton 1991 refers to as the 'chemosensitive area of the respiratory center') and the peripheral chemoreceptors in the arch of the aorta and the carotid bodies. So as well as looking at *Rutishauser 1994, Fig. 11.3A*, refer to *Fig. 12.6A* as well.

You should show how oxygen lack stimulates the peripheral chemoreceptors only when PO_2 levels have fallen considerably; and contrast this with the effect of only a slight rise in PCO_2.

You should establish how pH changes affect both the peripheral and the central chemoreceptors.

34. If I consciously increase my rate and depth of breathing, what effect will this have on my PO_2 and PCO_2 (assuming that I am in normal health at the time)? Will:

   - my PCO_2 fall
   - my PO_2 rise
   - both of these?

COMMENTARY ON QUESTION 34

If my respiratory health is normal, approximately what percentage oxygen saturation of the haemoglobin in my red blood cells will there be, during normal quiet breathing? If you remember the answer to this, it may help you with the correct answer to the second part of this question. Another help might be the comparison of the oxygen and carbon dioxide dissociation curves. See *Rutishauser 1994, p. 229*.

35. The peripheral chemoreceptors monitor the blood's PO_2 via the concentration of oxygen in the interstitial fluid bathing the receptor cells – the oxygen concentration being derived from the PO_2 of the blood. The peripheral receptors do not monitor the total oxygen content of the blood. Distinguish between these two measures of oxygen (see *Rutishauser 1994, p. 215*) and state the significance in the case of a person with anaemia.

You may now like to read about the control of respiration in different texts, just to check your understanding of this complex matter. See: *Hubbard & Mechan 1987, pp. 161–165; Jennett 1989, pp. 221–232* (very complex, so first try the 'Summary' on *pp. 231–232*); *Marieb 1992, pp. 754–759* (includes some excellent diagrams); *Guyton 1991, pp. 446–450*, 'Chemical control of respiration' (with separate sections for central and peripheral receptors).

Don't forget, too, the work you may have completed earlier in this Workbook on cardiac output and the blood – these factors are also very important in the control of respiration.

FURTHER STUDY

We well know from experience that exercise leads to us breathing faster and more deeply. We also know that at the end of a race, say, we continue to breath more deeply than usual – until we have 'caught our breath'. It is interesting to find out what effect exercise has on the physiology of breathing.

How much does the body's demand for oxygen rise? How much more carbon dioxide is created during hard exercise? Why do we continue to breath quickly even when we have finished running? Athletes talk of an 'oxygen debt' – what is this?

Why is exercise at a high altitude – as in climbing – more tiring than at sea level? Why do climbers on Everest use oxygen cylinders (except for the occasional oxygen-free attempt on the summit in recent years)?

Read the following texts on exercise and respiration: *Guyton 1991, pp. 945–947*, also *Ch. 43* on high altitude and space conditions; *Marieb 1992, pp. 759–760*, on the effects of both exercise and high altitude on the respiratory system.

 ACTIVITY

Get a colleague to check your pulse and respiratory rate and record them on a chart. (Try not to alter your breathing rate while it's being measured.) Then perform a suitable exercise, such as running a set distance or running up stairs. Have your pulse and respiratory rate checked and recorded when you've finished. See how long it takes for your respiratory rate to return to its normal level.

Change with your colleague so that he or she now does the energetic part of this activity while you carry out the recordings. Is there a marked difference between the time each of you took for your beathing to return to normal?

This activity might be made more interesting by including one of your colleagues who exercises regularly, and another who doesn't exercise or who is a smoker. But please don't choose an activity that might damage you.

■ The respiratory system (check quiz)

1. List the factors involved in getting oxygen from the lungs to the cells of the body. Here are three possible factors to start you off:

    - good cardiac output
    - adequate respiratory rate
    - lack of tissue oedema.

    Incidentally, how does oedema interfere with the supply of oxygen (and other nutrients) to tissue cells?

2. In severe anaemia, a person's oxygen-carrying capacity is greatly reduced – why?

    Discuss with your colleagues what might be the nursing care for an anaemic patient admitted to a medical ward. Don't worry about the medical treatment for the causes of the anaemia (such as stopping a chronic bleed, or making good dietary insufficiency). Would oxygen therapy play a part in your care?

3. Explain the interaction between perfusion and ventilation. Can you think of a situation where a person's perfusion is poor, though his ventilation is good? And another where the opposite is the case – perfusion is adequate but ventilation is poor.

    Explain, as if to a junior colleague, what is meant by the term shunt.

4. In cardiac resuscitation are you assisting ventilation, perfusion, or both?

5. Talking of resuscitation, here is something that puzzled me greatly as a student. Perhaps you'll be able to work out an appropriate answer.

    Imagine a first aider is performing cardiac massage and mouth to mouth resuscitation on someone who has collapsed. The ambulance paramedics will be able to pass a 'breathing tube' (endotracheal tube) and give the victim a high concentration of oxygen, but, until they arrive, all the first aider can do is perform mouth to mouth resuscitation.

    My problem was this: what is the point of doing mouth to mouth breathing, since all the patient receives is expired air with a relatively low PO_2 and a relatively high PCO_2? How can this help him physiologically?

6. Everyone knows – and most people admit – that cigarette smoking is bad for you. However, as a future health care professional you need to know rather more exactly how 'bad' smoking is for a person.

    Can you explain how smoking just one cigarette can affect the physiology of respiration? For example, what might happen to the smoker's PO_2, PCO_2, blood pH, his tidal volume, and so on?

    Now for the difficult part. Having assembled your answers to the above question, discuss with your fellow students how you'd adapt those answers for, say, a 12-year-old girl who had asked why cigarette smoking was bad for her.

7. You are shortly to admit a patient with a severe chest infection to your ward. Your mentor tells you that the patient is very breathless, frightened, and is coughing up a lot of sputum. She asks you to collect all the equipment needed for admitting this patient, and to give reasons for your choice.

    Discuss this nursing problem with your colleagues. Think about:

    - the patient's position in bed
    - observations you'll make on him
    - how to help his 'bad chest'
    - how to help his breathlessness

- how you can help ease his anxiety
- what medical tests he might need
- the patient documentation you'll need
- his ability to perform various activities of living.

Food and water

Contents

UNIT 4

■ An overview of digestion and excretion (work sheet)

Time for completion About 4 to 5 hours (no more than 1 'college day')

Overall aim To review the overall structure of the digestive tract; and to examine the mechanical breakdown of foodstuffs and the processes of mixing, peristalsis and defaecation.

Introduction There are at least two ways of dividing the digestive system for the purpose of study. One is to work your way down from beginning (the mouth) to end (the rectum and anus) dealing with each part of the system in turn. This is logical in that you study the physiology of digestion in the same order as pieces of ingested food are dealt with in the digestive tract.

Another way – and this is the method preferred by Rutishauser (1994) and Guyton (1984) – describes the digestive system from two angles: first, muscular movements within the system (e.g. mixing and peristalsis) and then chemical digestion (e.g. the secretion of enzymes by various organs of the system). In this Unit I have chosen to adopt the second approach. This Work Sheet concentrates on achieving an overview of the whole digestive system first, then moves on to looking at, in some detail, the mechanical or muscular aspects of digestion. Enzyme secretion, absorption, and control of these events, are covered in the following Guided Study.

My justification for choosing this approach is based on past experience of students' problems with studying the digestive system. Often, because of the relative complexity and detail of the various digestive enzymes, a question such as 'Explain how proteins are digested' tends to provoke an immediate examination of which enzymes are involved and what is the end product of their actions. In other words, the preliminary – but very important – mechanical processes of digestion have been forgotten. Studying this Work Sheet will, I hope, make such an omission impossible.

Background questions (This preliminary section is somewhat longer than in most Work Sheets, so I suggest you allow at least an hour for its study.)

1. The digestive system (otherwise known as the gastrointestinal or GI tract) can be described as a long, muscular tube. This tube varies in size of lumen (its diameter) and is open at both ends (mouth and anus). Using your preferred physiology textbook, draw a diagram showing the whole of this tube. Your diagram should include labels indicating:

   - the different parts of the small intestine
   - the different parts of the large intestine or colon.

   See, for example: *Wilson 1990, Fig. 9.1; Rutishauser 1994, Figs 6.1 and 6.5; Mackenna & Callander 1990, p. 64; Guyton 1984, Fig. 30.1.*
   Find out how long both the small and large intestines are. Whereabouts is the appendix in your diagram?

2. Add to your diagram the accessory organs of digestion. Include:

   - the salivary glands
   - the pancreas
   - the liver and gall bladder.

3. Copy a diagram from your chosen textbook of a cross-section of the digestive tract showing its lining and muscle layers. See, for example: *Wilson 1990,*

Figs 9.2, 9.25, and 9.34; Guyton 1984, Fig. 30.2; Mackenna & Callander 1990, p. 79; Rutishauser 1994, Fig. 6.5.

4. Draw a diagram showing the position of the peritoneum. Make sure you demonstrate its layers, and the fluid between them. Write a brief paragraph describing its position in relation to the major digestive structures, and its functions. See: *Wilson 1990, pp. 162–163, and Fig. 9.3; Rutishauser 1994, p. 110 and Fig. 6.4; Marieb 1992, Fig. 24.19.*

(Neither Guyton 1984 nor Guyton 1991 provides a useful reference to the peritoneum.)

 FURTHER STUDY

Find out the meaning of the term peritonitis. How does this condition occur and what are its clinical features? Stick to simple, basic descriptions at this stage of your training, and use them to help your understanding of the structure and functions of this important membranous structure.

5. Note the arterial blood supply to the digestive system, and the venous return from, in particular, the stomach, small bowel and large bowel to the liver. Don't go into much detail now, but simply note how nutrients, absorbed from the small intestine, pass to the liver along the hepatic portal vein. See *Wilson 1990, Fig. 9.8.* Where does this nutrient-rich venous blood flow to from the liver?

6. Describe, briefly, the autonomic nerve supply to the digestive tract. Look out especially for the vagus nerve which supplies, among other organs, the stomach and small intestine. (Where else in this Workbook have you met the vagus nerve?)

 COMMENTARY ON QUESTION 6

If you're easily confused by the two divisions of the autonomic nervous system – sympathetic and parasympathetic – complete the brief chart below by deleting the verb that does not apply:

- Parasympathetic – slows/hastens intestinal movements
- Parasympathetic – slows/helps secretion of digestive juices
- Sympathetic – slows/hastens intestinal movements
- Sympathetic – slows/hastens secretion of digestive juices.

Just think under which circumstances you are likelier to digest a meal properly:

- when you're rushing to get ready for work, or are anxious about exam results (fright, fight or flight; i.e. the sympathetic system predominates)
- when you're relaxed and contented, and the exam results have been excellent (i.e. when the parasympathetic system predominates).

Both sympathetic and parasympathetic nerves are organised into two nerve plexi (sing. plexus) found between layers of the digestive tract (see Question 3). Make notes on the submucosal plexus and the myenteric plexus. Where is each to be found, and what actions does each have on the digestive tract? See: *Wilson 1990, p. 165; Marieb 1992, pp. 772–774; Rutishauser 1994, p. 111; Hubbard & Mechan 1987, pp. 179–180; Guyton 1984, pp. 489–490.*

Specific questions

CHEWING AND SWALLOWING (THE MOUTH AND OESOPHAGUS)
Time for completion: about three-quarters of an hour

1. The teeth play a vital role in the initial tearing and chewing of food. It's probably unnecessary to remember the names of the different teeth, but have a look at a diagram, such as *Figs 9.12–9.15* in *Wilson 1990*, and note which of the adult teeth you think are involved in tearing, and which in grinding or chewing.

2. It's not only the teeth that are important for breaking up food in the mouth. In *Wilson 1990, Fig. 18.1*, or *Rutishauser 1994, Fig. 29.18*, you'll see the muscles of the jaw. There's also the tongue which helps to position food against the teeth. What fluid is secreted into the mouth during chewing to aid mixing and, eventually, swallowing? What substances are found in this fluid and what are their other functions? What other circumstances besides food actually in the mouth cause this fluid to be secreted? (This is another situation where physiology and psychology are closely linked.)

   You'll need to refer to *Ch. 29* in *Rutishauser 1994* (rather than Ch. 6) for matters pertaining to chewing.

 FURTHER STUDY

In this Work Sheet we don't study the structure of the teeth themselves, but this may be a useful subject for further study. There are obvious connections between it and both health education (oral hygiene) and the physiology of pain (the familiar experience of toothache).

---

3. Read about the structure of the oesophagus (American spelling: esophagus) and, in particular, the different types of muscle found in the upper and lower parts of this structure. Note how the oesophagus – which is in effect a muscular tube – leaves the laryngopharynx, passes behind the trachea (windpipe) and enters the stomach just below the diaphragm (see *Rutishauser 1994, Figs 6.6 and 29.9*).

   How is food propelled along this tube? (See *Rutishauser 1994, Fig. 6.7.*) Do you think it is possible to swallow food when lying down?

   Make notes, aided by a diagram if you wish, on the process of peristalsis. We'll come across this muscular progression in other parts of the GI tract.

4. How is the windpipe protected during the swallowing of food? This should be revision for you if you've already studied Unit 3.

   In health, what stops stomach contents from rising up into the oesophagus when we lie down after a meal?

FURTHER STUDY

Find out about a condition called hiatus hernia, perhaps illustrating it with a diagram. What is heartburn (which can be associated with a hiatus hernia, and which is nothing at all to do with the heart!)?

 Have a short break here.

FOOD MIXING IN THE STOMACH
Time for completion: about half an hour

5. First note the names given to the different areas of the stomach – fundus, body, cardia, etc. This will be very useful when you come to read medical notes about, for example, operations on patients with peptic ulcers. Then find out about the structure of the muscle layers in the stomach. What sort of muscle is it? In what directions do the various muscle fibres run?

   How does the stomach mix foodstuffs? Describe the contractions of the stomach muscles.

6. What are the functions of the cardia and the pylorus? What special structure has the pylorus, and how does this help its function?

7. What role is played by the vagus nerve in stomach mixing movements? When a patient with a peptic ulcer has a vagotomy (cutting of the vagus nerve) he tends to feel very full after only a small meal – why should this be so?

Remember that we'll consider secretion of digestive juices (which include gastric juice) in the following Guided Study, but the purpose of the stomach's mixing movements is to enable foodstuffs to be thoroughly mixed with gastric juice, thus aiding digestion. What is the name given to this semi-liquid mixture of food and digestive juices?

SEGMENTATION AND PERISTALSIS IN THE SMALL BOWEL
Time for completion: between a half and three-quarters of an hour

Within the small bowel, food is mixed with further digestive juices (from the pancreas, gall bladder, and small bowel itself) and then propelled along the substantial length of the small bowel to the large bowel. These mixing and propelling movements are different in nature.

8. The term given to mixing movements within the small bowel is segmentation. Make notes on this and show by means of diagrams how it works. Which smooth muscles in the gut wall are at work here, and which nerves are involved? Add notes to your diagram(s) showing that you understand the purpose of this segmentation. See: *Rutishauser 1994, Fig. 6.10; Mackenna & Callander 1990, p. 81; Hubbard & Mechan 1987, pp. 197–198; Guyton 1984, p. 494.*

9. How are the small bowel contents moved along the length of the gut towards the caecum (the first part of the large bowel)? Again, make notes on this

propulsive movement, using a diagram to show the difference between propulsion and segmentation. (References are similar to those given above.)

How long do the contents usually stay within the small bowel? (In the following Guided Study you'll discover that it is in the small bowel that much of the absorption of nutrients occurs; this explains the extended length of time that foodstuffs stay in the small bowel.)

 Have a short break here.

THE LARGE BOWEL AND DEFAECATION
Time for completion: up to 1 hour

10. You already have a diagram showing the various parts of the large bowel – ascending, transverse, descending, etc. Have a look at this diagram, which should demonstrate that the colon is not a straight tube, but one that is puckered into a series of shallow sacs (expanded areas). Check your own diagram with, for example, *Wilson 1990, Fig. 9.32,* or *Rutishauser 1994, Fig. 6.11.* Have you shown the caecum, and the appendix? (Strange how such a tiny structure can cause so much pain, as in acute appendicitis!)
    Now draw another diagram showing a cross-section of the large bowel, in particular its bands of longitudinal muscle fibres called taeniae coli.

11. Make notes describing the mixing movements of the colon. How long do foodstuffs usually remain in the colon? What is the main function of the colon, apart from acting as temporary storage for food remnants and as a means of expelling them?

 COMMENTARY ON QUESTION 11

Find out how big a volume of digested foodstuffs enters the colon each day, on average; then find out the volume of faecal matter expelled. How has the first volume been reduced to the second? Think about what happens, physiologically, when a person refrains from defaecation (perhaps because he is a patient on a ward for the first time and is embarrassed about having his bowels open in a comparatively public setting).

---

12. Now make notes on the process of defaecation itself (American spelling: defecation). You'll need to read about 'mass movements' within the colon, the nerve supply to the smooth muscles of the colon, and the function of the internal and external anal sphincters. How do the nerve supplies to the external and internal sphincters differ? How do voluntary muscles assist in the process of defaecation? See: *Wilson 1990, pp. 181–183; Rutishauser 1994, p. 115 and Fig. 6.13; Guyton 1984, p. 496 and Fig. 30.8; Mackenna & Callander 1990, pp. 86–88; Hubbard & Mechan 1987, pp. 215–216; Marieb 1992, pp. 806–808.*

13. What is the gastrocolic reflex? When does it occur and what is its connection with defaecation?

FURTHER STUDY

We've all read about the benefits of a high fibre diet. It may be interesting to discover how such a diet actually helps the process of defaecation. Does a high fibre meal stay within the GI tract for more or less time than a low fibre meal? Is defaecation made easier with a high fibre meal?

Fibre will be mentioned in the Suggested Reading on nutrition a little later in this Workbook.

VOMITING
Time for completion: about a quarter of an hour

14. Finally we look at the mechanical actions involved in vomiting, because this is related to peristalsis which you've already described.

Find out what happens to the pylorus and cardia of the stomach during vomiting, and the strong actions of the stomach's muscles and the diaphragm. See: *Mackenna & Callander 1990, p. 73; Marieb 1992, p. 792; Guyton 1984, p. 496.*

Make brief notes on the causes of vomiting, and the nerve pathways involved.

Further exploration of the digestive system (guided study)

Time for completion   About 8 to 10 hours (or no more than 2 'college days')

Overall aim   To consider the chemical digestion of foodstuffs, their absorption from the gut, their utilisation and storage, and the hormonal control of digestive system activity.

Introduction   You have already studied how food is physically broken down within the GI tract. This important part of the digestive process is a preliminary to chemical digestion, whereby food is mixed with substances (many of them enzymes) secreted by various organs of the tract, such as the liver and pancreas.

To remember the names of all these enzymes is perhaps a daunting task. So try to finish this Guided Study with a working knowledge of whereabouts in the GI tract each of the main types of food – proteins, fats and carbohydrates – is broken down chemically; the end-product of the digestive process for each type of food; and one or two examples of the enzymes involved.

While working on this Guided Study, you may find it useful to construct one or two simple diagrams which will help you remember where digestive processes occur for certain foods. I will be providing suggestions for such diagrams.

Background questions   1. Look back at the original diagram you drew showing the whole of the digestive system 'tube', together with the accessory organs. Make sure you understand how each of the accessory organs delivers its secretions to the GI tract; for example, how are the pancreas and gall bladder connected to the small intestine?

2. What is the definition of an enzyme? Have you come across enzymes elsewhere in human physiology?

Specific questions   **SECTION 1: CHEMICAL BREAKDOWN OF FOOD**
Time for completion: about 3 to 4 hours (i.e. a whole morning or afternoon)

The mouth

1. Saliva is produced from pairs of glands. Check that you know how many pairs, their names, and where they are situated in the mouth.

2. What are the constituents of saliva? Is there an enzyme in saliva, and on which of the three main foodstuffs does it work?

3. From the preceding Work Sheet you'll have discovered what controls the flow of saliva. What effect does chewing have on its production? Can you discover any drugs that reduce saliva flow? When are such drugs used therapeutically?

 COMMENTARY ON QUESTION 3

Perhaps you have been a surgical patient, and been given a premedication injection. One of its functions is to make you sleepy and less anxious prior to anaesthesia. As well as this, you may have noticed that your mouth felt dry. (This could have been the action of the sympathetic nervous system as well as the drug.) Why do you think the anaesthetist wants to dry your salivary secretions during operation?

For Questions 1 to 3, see: *Wilson 1990, pp. 171–172; Hubbard & Mechan 1987, pp. 186–187; Guyton 1984, pp. 497–498; Rutishauser 1994, p. 117; Marieb 1992, pp. 777–779.*

FURTHER STUDY

Briefly, make notes on other functions of saliva besides its digestive ones. What effect, for example, does saliva have on one's speech? (Have you ever tried to speak in public, for example to your whole class? Is it easier or harder to speak when your mouth is dry?)

The stomach

4. The mixing and emptying movements of the stomach have already been studied. As a form of revision, make a drawing similar to Figure 4.1, and complete the labels showing the various parts of the stomach.

   Add the diaphragm to your drawing.

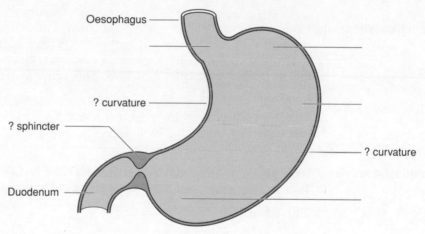

Fig. 4.1 Parts of the stomach.

5. Gastric glands are found throughout the stomach lining. Different cells within these glands produce different substances. Which are the cells that produce:

   - hydrochloric acid (HCl)
   - digestive enzymes
   - mucus?

   What is the action of mucus? (Incidentally, note the different spelling of two very similar words: mucus, the noun, meaning the substance produced; and the adjective, mucous, as in mucous membrane or mucous lining.)

6. Find out the names of the gastric enzymes and their actions on the three main foodstuffs: carbohydrates, proteins, and fats. What is the action of hydrochloric acid?

   One of the enzymes you've listed is secreted in its non-active form as pepsinogen. How is this substance converted to its active form, and what is it then called? Why do you think it is initially secreted in an inactive form? (Think about the foodstuff it acts on, and the composition of stomach tissue.)

7. How is the production of gastric juice controlled by the vagus nerve? (We'll return to the subject of control of secretion in Section 3 of this Guided Study.)

8. What effect do the following have on gastric juice and mucus production:

   - cigarette smoking
   - alcohol (especially spirits)
   - strongly spiced foods?

FURTHER STUDY

Aspirin is a commonly used drug which can have serious effects on the stomach lining if taken in excess. Find out about the action of this drug on the stomach, the clinical features you might expect from too high an intake of aspirin, and the basic advice you might give a patient who is taking aspirin. For example, should aspirin be taken before or after meals, on an empty or full stomach?

Another drug, which is designed to block HCl production from the stomach lining, is cimetidine. Ranitidine is a later version. Find out how this drug works, and its effect on a substance called histamine. You may have heard of antacids such as Milk of Magnesia, which you can buy without a prescription. How do cimetidine and ranitidine differ from these antacids?

---

9. This is perhaps an appropriate place to begin a diagram that provides an overview of the action of digestive enzymes on the three main groups of food. The diagram – of which Figure 4.2 provides the outline – is intended to be not too detailed, but rather a summary to provide 'information at a glance'.

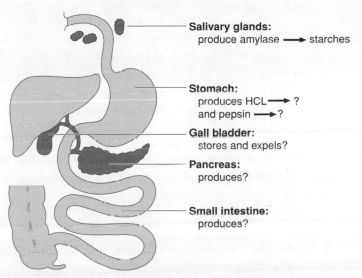

Salivary glands:
produce amylase ⟶ starches

Stomach:
produces HCL ⟶ ?
and pepsin ⟶ ?

Gall bladder:
stores and expels?

Pancreas:
produces?

Small intestine:
produces?

Fig. 4.2 Representational overview of digestive juices and their actions. Add the missing information to your diagram. → = acts on.

For fuller detail, you could build up your own digestive activity chart, showing not just the enzymes and the foods they work on, but how such foodstuffs are gradually broken down into their final, absorbable state. You might refer to *Wilson 1990, pp. 192–193*, or *Rutishauser 1994, Table 6.2*, but don't just copy the information from there. You should preferably extract from your chosen texts the relevant information for the gradual construction of your own chart.

For Questions 4 to 9 see: *Wilson 1990, pp. 173–177; Rutishauser 1994, p. 119; Hubbard & Mechan 1987, pp. 191–193; Guyton 1984, pp. 498–500; Marieb 1992, pp. 784–789.* (Note that much of the material in Marieb relating to the stomach, refers to neural and hormonal control of gastric activity. This reference is probably better left to the final part of this Guided Study.)

Find out the average daily volume of digestive juice secreted from the salivary glands and stomach. You may be , rprised at the answer. Begin a chart listing the digestive organs and the volume secreted by them daily.

 Take a break here.

The pancreas

10. First ensure that you can describe the position of the pancreas in relation to:

 - the stomach
 - the duodenum
 - the liver and gall bladder.

 How is the pancreas connected to the small bowel? (In other words, how does pancreatic juice flow from the pancreas to the duodenum?) Is there any relationship between the pancreatic duct and the duct from the gall bladder?

11. Find out the digestive enzymes secreted by the pancreas, enter them on your digestive activity chart, and describe how each enzyme works on its appropriate foodstuff (carbohydrate, protein or fat). As in the stomach, these enzymes are initially secreted in an inactive form – why is this? How are they activated?

12. The pancreas also secretes sodium bicarbonate (sodium hydrogen carbonate). What is the function of this highly alkaline substance?

 What pancreatic cells secrete sodium hydrogen carbonate? Are these the same cells that secrete the digestive enzymes?

 It is perhaps worth noting that vagus nerve stimulation also causes pancreatic juice secretion, but it is far less important than the influence of hormones.

For Questions 10 to 12 see: *Wilson 1990, pp. 179 and 183; Rutishauser 1994, p. 119; Hubbard & Mechan 1987, pp. 199–201; Guyton 1984, pp. 500–501; Marieb 1992, pp. 801–802.*

The liver, and production of bile

13. The production of bile is one of many functions of the liver. In this Workbook

we don't study the liver as a separate entity, so it might be worth reading *Wilson 1990, p. 186*, for a useful overview of liver functions. Note how important the liver is in connection with the excretion of drugs.

14. Now describe the position and structure of the gall bladder, noting how it is connected with both the liver and the small bowel. Draw a diagram of the gall bladder and label the various ducts.

15. What are the constituents of bile, and how is it formed? What are bile salts and what is their function? (Bile is not an enzyme, so what does it actually do to foodstuffs with which it is mixed?)

16. What are the functions of the gall bladder? Bile is manufactured in the liver, but what does the gall bladder do to this bile before it is secreted into the duodenum?

    Again bear in mind that we'll be studying the control of digestive juices (including bile) later in this Guided Study.

For Questions 13 to 16 see: *Wilson 1990, pp. 186–187; Rutishauser 1994, pp. 120 and 183; Hubbard & Mechan 1987, pp. 205–209; Guyton 1984, pp. 501–502; Marieb 1992, pp. 798–800.*

The small intestine

It is in the small intestine that foodstuffs are finally digested, and much of the absorption occurs – hence the great length of this part of the GI tract. Pancreatic juice and bile contribute greatly to this digestive process, but secretions from the small intestine itself also play a part.

17. First note how the lining of the small intestine greatly increases its surface area. Your notes should mention:

    - circular folds
    - villi
    - microvilli.

    Draw a digram showing the villi, together with their blood and lymphatic networks, and their epithelial cell walls. (See, for example, *Wilson 1990, Fig. 9.30.*)

18. Make notes on how small intestinal juice is formed and secreted into the lumen of the bowel, in a different manner from the production of, for example, pancreatic juice. Note the production of mucus as well as enzymes. Add these enzymes to your digestive activity chart, showing on which foodstuffs they work, and what is the result of their digestive processes.

For Questions 17 and 18 refer to: *Wilson 1990, pp. 177–180; Rutishauser 1994, p. 121; Hubbard & Mechan 1987, pp. 198–199; Guyton 1984, p. 503; Marieb 1992, pp. 792–796 and 802.*

Finally, before you close your books, check your digestive activity chart for clarity, then complete the digestive juices volume chart you began earlier. Add to the latter chart the average daily volume of digestive juices secreted by the pancreas, liver (via the gall bladder) and the small intestine. What is the total volume secreted each day? (Just for interest, convert this total into, say, the number of cups.) What happens to all this fluid?

 This is a good opportunity for a lengthy break. Your studies so far should have taken up a whole morning or afternoon.

 FURTHER STUDY

If you want to check out your knowledge of the digestive system with more detailed texts, try *Hinchliff & Montague 1988, Ch. 5.1*. This chapter also deals with the absorption of nutrients (which we have yet to study). For information on the liver you will need *Ch. 5.2*.

Also well worth dipping into is *Guyton 1991, Chs 62, 63 and 64*, with *Ch. 65* worth a glance after we've studied absorption of nutrients. But do make sure you have a good working knowledge first before trying this highly detailed text.

SECTION 2: ABSORPTION OF NUTRIENTS
Time for completion: 2 or 3 hours

19. The whole point of the mechanical and chemical breakdown of food is to bring it to such a condition that it can be absorbed through the gut wall into either the blood capillaries or lymphatics. First, then, read about the three methods of absorption, as described in *Rutishauser 1994, p. 125*:

    - passive diffusion
    - carrier-mediated transport
    - endocytosis and exocytosis.

    In some cases diffusion occurs when substances are broken down to so small a size that they can pass between the cells of the villi. Larger materials need to be absorbed by some other method, such as being carried across the cell membrane by a carrier protein. In some cases, cellular energy is expended in order to 'push' certain substances 'uphill' (i.e. against the concentration gradient).

20. Check your knowledge of the structure of the small intestine and, especially, its villi (sing. villus). Note the presence of tiny lymphatics within the villi as well as blood capillaries. See *Fig. 6.26* in *Rutishauser 1994*.

21. The main part of this Section is a study of how each of the major groups of nutrients is absorbed within the small intestine. Do use diagrams to help clarify your notes. You should study the absorption of:

    - monosaccharides
    - amino acids
    - fats
    - fat- and water-soluble vitamins
    - minerals, such as calcium, sodium and iron
    - water.

COMMENTARY ON QUESTION 21

It's easy to forget the absorption of water! But just look at the digestive juices volume chart that you've been maintaining throughout this Guided Study. Remember how you converted all those millilitres of digestive juices secreted into the gut to the equivalent cups of fluid? In health, the greater part of this fluid has to be reabsorbed by the gut. In some conditions, such as cholera, the gut both secretes fluid and is unable to reabsorb the usual volume of water, and severe diarrhoea results, often, in the Third World, leading to the death of the sufferer.

You may find the absorption of fats the most complicated. You'll need to understand the formation and function of micelles: see *Fig. 6.27* in *Rutishauser 1994*. You'll find that fats are absorbed first into the lymphatics within the villi of the small intestine.

The absorption of iron is interesting, in that it is affected by the pH of the bowel contents. Whereabouts in the small bowel is iron more readily absorbed?

My list of suggested reading is a little shorter than usual, because, for this subject, I'm advising you to leave Guyton 1984 alone. Try: *Rutishauser 1994, pp. 125–129; Hubbard & Mechan 1987, pp. 210–213; Mackenna & Callander 1990, pp. 82–84* (a good summary, but it's best not to use this as your main text); *Marieb 1992, pp. 813–815.*

22. Now we consider absorption within the large bowel, a much simpler affair than absorption in the small intestine. Note how water is absorbed in the colon largely following the absorption of ions such as sodium and chloride. See: *Rutishauser 1994, p. 129; Hubbard & Mechan 1987, pp. 216–217.*

Drugs can be given rectally. Some anti-inflammatory drugs (used in some forms of arthritis) are given as suppositories in an attempt (not altogether successful) to avoid their common side effect of irritating the gastric mucosa. You may like to discuss this problem with colleagues: how can an anti-inflammatory drug, given rectally, cause irritation to the stomach? It's easy to see how tablets that are swallowed can cause this – but suppositories?

23. The final part of this Section concerns the fate of digested food. Having eaten a meal, mechanically and chemically digested it, and then absorbed its nutrients, what happens to those nutrients once they have reached the bloodstream?

Wilson 1990, pp. 187–191 provides excellent background reading on metabolism. You can then back this up by reading *Rutishauser 1994, pp. 266–274*, and, in greater detail, *Marieb 1992, pp. 835–860* (many useful diagrams here).

Which nutrients provide energy quickly? What happens if we eat too much carbohydrate, protein or fat? How can glucose be made out of excess fat in the diet, and how can stored fat be converted into glucose to provide energy?

24. If you wish, you may like to read about ATP (adenosine triphosphate) which is a stored form of energy. You'll find this mentioned briefly in *Wilson 1990, pp. 187–191*, and on *pp. 41–42* of *Mackenna & Callander 1990*. Then read *Rutishauser 1994, p. 211*. How is ATP formed, and under what circumstances can it be split to release energy?

Note: Questions 23 and 24 may be completed after you have worked through the Suggested Reading on nutrition which follows this Guided Study.

SECTION 3: HORMONAL CONTROL OF DIGESTIVE SYSTEM ACTIVITY
Time for completion: 2 or 3 hours

Throughout this Guided Study you'll have noticed how digestive system activity is affected by the nervous system, for example the vagus nerve. You've seen how saliva can be produced not just by the mechanical action of chewing, but also by the thought or smell of food. Cutting of the vagus nerve supply to the stomach (vagotomy) causes a decrease in both the amount of acid produced and gastric motility. The former result is what is required when vagotomy is performed as treatment for peptic ulcers. Reduced gastric motility is one of the unwanted effects of the operation, in that the stomach empties more slowly; even after a few mouthfuls the person feels full.

For this final Section of the Guided Study we look at the hormonal (humoral) control of digestive system activity.

Control of stomach activity

25. The arrival of food in the stomach increases the secretion of a hormone called gastrin. Find out from where this hormone is produced, and what are its effects. For instance, does gastrin influence gastric juice production, or gastric motility, or both? See: *Wilson 1990, p. 176, and Fig. 9.26; Rutishauser 1994, p. 133 and Table 6.9; Jennett 1989, pp 251–253.*

Control of pancreas, gall bladder, and small intestine activity

Following thorough mixing in the stomach with gastric juice, chyme passes through the pylorus and enters the duodenum. The presence of chyme in the duodenum increases the production of a hormone called secretin by cells of the duodenal mucosa. (*Jennett 1989, p. 260,* provides an interesting history of secretin and its discovery.)

26. What is the effect of secretin production on pancreatic secretions and, therefore, on the duodenal contents? I don't think it's enough to say that the pancreas secretes pancreatic juice: there's something special about pancreatic secretions caused by the production of secretin.

27. What effect does secretin have on the production of bile in the liver?

28. Another hormone is secreted because of the arrival of chyme in the duodenum. For easy reference, this hormone is often referred to as CCK-PZ, but can you discover its full name? Where is it produced?

29. What effect does CCK-PZ have on the gall bladder?

30. What effect does CCK-PZ have on pancreatic secretions?

31. What effect does CCK-PZ have on gastric emptying?

32. Finally, if chyme arriving in the duodenum has a high fat content, another hormone is produced which is referred to as enterogastrone. What effect does it have on gastric secretions and motility? Read the brief notes provided by *Wilson 1990, p. 176,* and *Jennett 1989, p. 262,* about the name of this hormone.

References for this Section include: *Wilson 1990, pp. 179–180; Rutishauser 1994, p. 132; Jennett 1989, pp. 255–262.*

I have attempted to draw a diagram summarising the production and actions of gastrin, secretin and CCK-PZ. See what you think of Figure 4.3 – perhaps you can improve it?

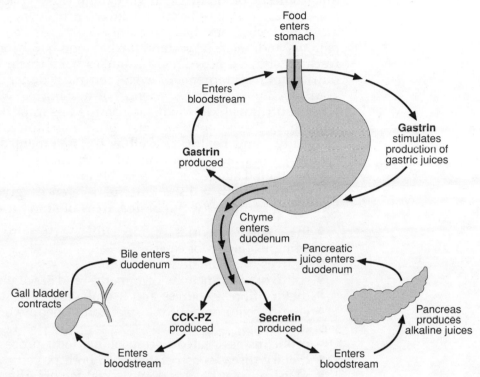

Fig. 4.3 Summary of hormonal control of the digestive system.

Your own notes and diagrams may be checked against those in *Mackenna & Callander 1990, pp. 71, 75, and 77; p. 88* provides a useful summary of the nervous control of gut movements.

■ Nutrition in health and illness (suggested reading)

Nutrition has been, in the recent past, one of the forgotten areas of nursing. Presenting food to patients became, about 20 years ago, a non-nursing duty so that domestic or kitchen staff gave out meals to patients on their wards. While nurses were certainly better employed in delivering bedside care, the situation arose whereby they had no idea what food was being given to their own patients and, more importantly, what food had been eaten. Small wonder that research showed how ignorant nurses were (including trained nurses) of the nutritional requirements of even their most dependent patients (Jones 1975).

My feeling is that the 'science' of nursing is swinging back to a situation where nutritional knowledge is regarded as important to the body of nursing knowledge. This may be because good nutrition is a vital part of the concept of health – and health, as well as the treatment of ill health, has become a central part of the modern nursing curriculum.

First, then, I suggest some possible areas of importance to your studies, and follow them with a list of useful references given, as usual, in order of increasing complexity. Suggested areas of study are:

- the main categories of foodstuffs – carbohydrates, proteins, and fats, providing examples of each
- sources and functions of vitamins
- sources and functions of minerals and trace elements
- dietary fibre – sources and usefulness
- energy requirement for a selection of people (e.g. child, adolescent, office worker, nurse)
- problems associated with cooking, food processing and storage
- dietary diseases associated with both poverty and affluence
- physical, psychological and social factors affecting diet (e.g. gut surgery, depression, size of family)
- the effectiveness of slimming diets.

Choosing, for example, a 15-year-old boy and a 19-year-old student nurse, construct for each a day's menus designed to meet their nutritional needs. Are there any 'danger areas' associated with their food intake (e.g. not enough time to eat properly, preference for 'junk' food)?

Suggested reading

Wilson 1990, Ch. 8, 'Essential nutrients'
Roper, Logan & Tierney 1990, Ch. 9, 'Eating and drinking'
Boore, Champion, & Ferguson 1987, Ch. 14, 'Dietary therapy and nutritional care' (see especially *pp. 332–341*, on nutritional problems of the hospitalised patient)
Mackenna & Callender 1990, pp. 32–36
Marieb 1992, Ch. 25, and especially *pp. 823–834*
Rutishauser 1994, Ch. 14
Guyton 1984, Ch. 32, especially *pp. 530–537*
Guyton 1991, Ch. 71 (use this as a means of checking information gained from other books, not as your principal text).

REFERENCES
Boore J, Champion R, Ferguson M 1987 Nursing the physically ill adult. Churchill Livingstone, Edinburgh
Jones D 1975 Food for thought. Royal College of Nursing, London
Roper N, Logan W, Tierney A 1990 The elements of nursing. Churchill Livingstone, Edinburgh

■ The digestive system (check quiz)

1. What foods provide you with a *quick* supply of energy? For example, you are about to go on a rock climb which might take you about 2 or 3 hours. There are four stances (places to stop and rest) along the route. What foods might you pack in your rucksack to help you regain energy and strength for the remainder of this strenuous climb?

2. Now decide which foods might be best for *prolonged periods* of work. For example, what should you eat before going on duty on a busy surgical ward for an 8-hour shift? (You could always take along items of food for a quick snack during your spell on duty – and these could well be the same foodstuffs taken on your rock climb, as above.)

3. What are the benefits of a high fibre diet? How much fibre should there be in the daily diet of a healthy young adult, and in which foodstuffs will this amount of fibre be found?

4. You have decided that you are about 5 kilograms overweight, and so you plan to go on a diet. Bearing in mind that you're still working on that busy surgical ward, how might you plan your diet so as to help you lose weight effectively, but to keep up your strength as well. (I'm going to disallow 'miracle diets' such as are found in magazines. You must stick to ordinary foods that are available daily. Think about the amount of various foods you might eat, how satisfying each meal might feel, and how frequently you plan to eat throughout the day. Also, what part might exercise play in your diet?)

5. Forget about your diet now; in front of you is a bacon sandwich consisting of lean, grilled bacon, two slices of wholemeal bread, and low fat spread. Explain, step by step, the mechanical and chemical breakdown (digestion) of each of the principal foodstuffs in the sandwich – protein, carbohydrate, and fat. Use diagrams if you feel they'll help your explanations. What other important nutritional elements are present?

6. How is the small intestine constructed to aid absorption of nutrients? Describe in detail the absorption of fats. What part do lymphatic vessels play in this process?

7. Explain the gastrocolic reflex. Jill has a fractured spine as a result of a car accident, and she has no feeling below the level of her waist. One of her most difficult problems to deal with is faecal incontinence. Could you suggest any way in which the gastrocolic reflex could be utilised to help counteract this incontinence? (Bear in mind that although nerve impulses are unable to reach her brain because of the severed spinal cord, spinal reflexes – below the level of the injury – are still possible.)

8. Both patients and student nurses sometimes complain of an alteration in their bowel habits while in hospital. Explain the physiological reasons for this in:

   a. a middle-aged patient following a heart attack who is presently being nursed in bed
   b. a student nurse whose final examinations are fast approaching.

   One of these individuals complains of increased frequency of bowel movements; the other of decreased frequency. Which person do you think would complain of which alteration?

■ An overview of the urinary system (work sheet)

Time for completion	About 4 to 5 hours
Overall aim	To achieve a basic understanding of the structure and function of the urinary system, including the production of urine and micturition.
Introduction	The kidneys filter approximately 150 litres of fluid a day yet, in health, only about 1 or 2 litres of urine are excreted from the body. In the urine there are substances that are either unwanted by the body or, in large amounts, are actually dangerous. Also, when we take drugs such as analgesics or antibiotics they are often excreted in the urine. (A patient on an antibiotic such as ampicillin, for example, produces urine that smells strongly of the drug.)

Urine production occurs under greatly changing physiological circumstances: in rest or exercise, in hot or cold climates, when we are drinking copiously or are fasting (e.g. before an operation). The aim of this Work Sheet and the following Guided Study is to help you understand how kidney function adapts to such changes in physiological conditions. |

Background questions

1. Draw and label a diagram showing the urinary (renal) system. It should include the left and right kidneys, the ureters, bladder, and urethra.

2. Make notes on how the kidneys are partially protected from trauma and from cold. Are there any body structures that act to protect the kidneys?

3. Describe the blood supply to the kidneys, noting what percentage of cardiac output in health supplies the kidneys. How is blood returned to the general circulation from the kidneys?

4. What is the position of the kidneys in relation to the peritoneum and other abdominal organs? This is important when, later in your nurse training, you consider operations on the kidneys.

5. Finally, list the normal constituents of urine. This will help you follow the various stages of urine production later on. Can you group these various substances under subheadings in order to help you remember them?

Specific questions

FILTERING THE BLOOD
Time for completion: about 1½ hours

1. Draw and label a diagram of a split or coronally sectioned kidney. See for example: *Wilson 1990, Fig. 10.4; Rutishauser 1994, Fig. 8.3A.*
   See also *Marieb 1992, Fig. 26.3,* for a photograph of a split kidney. Try to relate the different areas of kidney tissue you see there to those on your own diagram.

2. Draw and label a diagram of a nephron, the functional unit of the kidneys. Make sure you include the blood capillaries as well as the tubules. See: *Wilson 1990, Fig. 10.5; Rutishauser 1994, Fig. 8.2A; Marieb 1992, Fig. 26.4.*
   Note how the cells of the nephrons differ in structure according to where in the nephron they're situated. For example, look at the cells making up the loop of Henle. Approximately how many nephrons are there in each kidney?

3. Draw another diagram showing how a nephron is positioned in relation to the areas of a split kidney. Where, for example, are the glomerulus, the loop of Henle, and the collecting tubule, in relation to the cortex and medulla of the kidney? See *Rutishauser 1994, Fig. 8.3B.*

4. Draw a detailed diagram of the glomerulus and Bowman's capsule. Label the afferent and efferent capillaries. *Rutishauser 1994* has several useful diagrams (*Figs 8.2B, and 8.5A & B*); *Guyton 1984* has an electron micrograph of a glomerulus (*Fig. 22.4*) showing the tiny foot processes and slits through which glomerular filtration occurs. Can you relate these features to, for example, *Fig. 8.5B* in *Rutishauser 1994*?

5. Make notes on how filtration – the first stage of urine formation – occurs in the glomerulus. Describe the different pressures which give rise to the glomerular filtrate, and show these pressures in diagram form. See: *Hubbard & Mechan 1987, pp. 96–97; Rutishauser 1994, Fig. 8.6, and p. 161; Guyton 1984, pp. 357–359; Marieb 1992, pp. 880–882; Jennett 1989, pp. 284–287.*

6. What factors can affect the net pressure in the glomerulus? See *Rutishauser 1994, Figs 8.6 and 8.7.*

7. Note the composition of the glomerular filtrate. What substances go through the glomerulus into the filtrate, and what substances, in health, do not pass through the filter? Figure 4.4 is my attempt to represent this in a highly diagrammatic form – perhaps you'll be able to improve on it. Complete your own version of the diagram by adding to the lists of substances that do and do not pass through the filter.

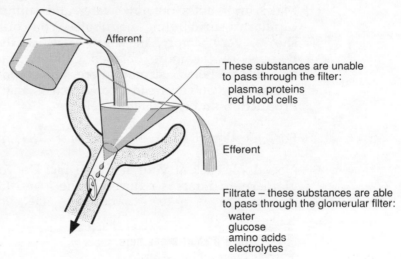

Afferent

These substances are unable to pass through the filter:
  plasma proteins
  red blood cells

Efferent

Filtrate – these substances are able to pass through the glomerular filter:
  water
  glucose
  amino acids
  electrolytes

Fig. 4.4 Glomerular filtration. Complete the lists in your diagram.

 FURTHER STUDY

Read about tests of kidney function which assess glomerular filtration. How is the glomerular filtration rate (GFR) measured and what is its significance? See, for example, *Blandy & Moors 1989, pp. 30–32.*

RECOVERING VITAL CONSTITUENTS FROM THE FILTRATE
Time for completion: 1 to 1½ hours

You'll have noticed earlier that the filtrate contains glucose, yet you will probably know that, in health, glucose does not appear in the urine. In this section of the Work Sheet, we'll look at how substances the body needs are recovered from the filtrate. Waste products or excess electrolytes, for example, are allowed to be excreted in the urine; but the body needs glucose for energy. It is too important to waste, and so it is reabsorbed back into the blood.

In order to illustrate the principle of selective reabsorption, we'll look here mainly at how glucose, amino acids, sodium and water are drawn from the filtrate back into the blood. Electrolyte and water balance will be considered in more detail in the Guided Study that follows. First, however, list those constituents of the filtrate that are reabsorbed.

8. From which parts of the nephron are these constituents reabsorbed:

   - sodium
   - glucose
   - amino acids
   - water?

9. By what process are glucose and amino acids reabsorbed?

10. Find out the normal blood glucose range, and what is meant by the renal threshold for blood glucose. (See *Rutishauser 1994, Fig. 8.8, and p. 163*.) You could also discover how a patient's urine is tested for glucose, and what the term is for glucose in the urine.

11. Make brief notes on how water and sodium are reabsorbed back into the capillaries surrounding the tubules. You could perhaps devise a diagram that shows, very simply, in which parts of the nephron reabsorption occurs. Because of the more detailed Guided Study that follows, please restrict yourself to short, basic notes here. You could, however, remind yourself what hormones you've already come across that influence the volume of urine production and electrolyte levels.

For Questions 8 to 11, see *Rutishauser 1994, pp. 163–169*.

Now look back at your notes so far. Do you understand how 150 litres of glomerular filtrate is reduced to the 1 or 2 litres of urine excreted each day?

 Have a short break here.

SECRETION AND ELIMINATION OF WASTE PRODUCTS
Time for completion: about half an hour

12. Creatinine and urea are waste substances that are excreted via the urine. First, find out where these two substances come from.

13. Now make notes on how they get into the kidney tubules and how they are finally excreted from the body. For a description of the fate of urea see: *Rutishauser 1994, p. 170; Guyton 1984, p. 361*.

Note that some substances remain in the tubule (and so are excreted in the urine) because of a failure in tubular reabsorption (i.e. failure to be reabsorbed back into the blood). Some substances are secreted from the capillaries into the tubular cells, and thence into the tubular fluid. This process is more selective than failure of reabsorption.

14. Find out how antibiotics are excreted via the urine.

TEMPORARY STORAGE OF URINE, AND MICTURITION
Time for completion: about 1 hour

15. In health urine is produced almost continually (though in differing amounts at different times of the day). Make notes on how urine is conveyed from the kidneys to the bladder. Describe the structure of the ureters and bladder. What sort of cells line the bladder? What is the trigone? See: *Rutishauser 1994, Fig. 8.16; Marieb 1992, pp. 894–896; Mackenna & Callander 1990, p. 161.*

16. Describe the internal and external sphincters of the bladder and their nerve control.

17. Describe the events leading up to micturition – what happens to the bladder wall as it fills with urine, how nerve 'messages' are passed to the brain, and how a person can, within limits, control the passing of urine until a convenient time. Note the normal comfortable capacity of the bladder. See: *Hubbard & Mechan 1987, p. 90; Rutishauser 1994, Figs 8.17 and 8.18; Marieb 1992, p. 897.*

18. Define the following commonly used terms:

    - diuresis
    - anuria
    - dysuria
    - polyuria
    - nocturia.

    In both anuria and retention of urine (the latter perhaps because of an enlarged prostate gland in elderly men) no urine is passed. But there's an important difference between the two terms – what is it? Under what circumstances do you think polyuria might occur?

 FURTHER STUDY

Incontinence of urine is a distressing occurrence which is probably more common than you realise. You should not associate incontinence with the elderly. There are many different forms, but one that you may find particularly interesting to study (because of its clear link with physiology) is stress incontinence. This especially affects women. Some nurses specialise, following registration, as incontinence advisors to patients living in the community.

Now look back at the list you made earlier of the constituents of urine. Apart from the electrolytes (such as sodium and potassium) which we're leaving till the following Guided Study, do you understand how each of those substances appears in the urine? Are you clear about the absence of glucose from your list?

REFERENCE
Blandy J, Moors J 1989 Urology for nurses. Blackwell Scientific Publications, Oxford

■ Fluid and electrolyte balance (guided study)

Time for completion
About 8 hours (or 1½ 'college days')

Overall aim
To build on the student's present knowledge of urine production by examining how, under varying physiological conditions, water and electrolytes are excreted or conserved; and to examine how the body's pH is maintained within normal limits.

Introduction
Here are two brief patient profiles which I think will illustrate the importance of the body remaining within its correct range of fluid and electrolyte levels.

The first patient is a 50-year-old lady with heart failure. (The reasons for this condition don't concern us here, but it's perhaps worth saying that her heart can no longer efficiently pump blood around her body.) She is breathless because her lungs contain fluid. Her legs are swollen with fluid, especially late in the evening. If you gently push your finger into the swollen (oedematous) limb, you'll see that the hollow doesn't fill immediately; instead a small pit remains in the waterlogged tissues. The patient finds the least exercise exhausting. Her doctors describe her as being in fluid overload.

My second patient is a young child living in poor home conditions. He has picked up a 'tummy bug' and has had diarrhoea for about 3 days now. His skin is dry and appears somehow loose, so that you can pinch a fold of skin on his arms away from the underlying fat and muscle. His eyes seem to be sunken and have lost their sparkle. Like my first patient, he finds every effort exhausting. He is described as being dehydrated.

These two patient profiles illustrate severe failure of fluid and electrolyte balance. In our own lives, we experience less extreme changes in our fluid balance. We might go to a party and drink 5 pints of beer in one evening; or we may find ourselves on restricted fluid intake because of a medical investigation or operation. Our bodies can cope with such physiological changes; the bodies of my original two patients cannot cope with their extreme changes.

Background questions
1. Find out the normal serum levels of sodium, potassium, chloride and calcium.

2. What is the total volume of fluid found within the adult body? In what form is this fluid found; for example, is it always in the form of a liquid, as in the plasma part of blood? See *Rutishauser 1994, p. 237.*

3. It's easy to understand how our daily intake of fluid occurs in the form of drinks. But water is also found, to varying degrees, in our food. You might immediately think of on orange or melon, both of which have an obvious water content. Can you list about five other foods that you think will contain a substantial amount of water, though not especially obviously?

Specific questions
SECTION 1: MOVEMENT OF ELECTROLYTES AND WATER
Time for completion: about 1 hour

1. Draw a diagram showing the fluid 'compartments' of the body. Show the proportion of total body fluid found in:

- the intracellular compartment
- the extracellular compartment
- the interstitial compartment

and give examples of each compartment. (For example, to which compartment does fluid found inside blood cells belong?)

2. These compartments aren't necessarily exclusive in that water and dissolved substances can sometimes move between them. Make notes and, if helpful for you, diagrams on diffusion and osmosis. In one, water is drawn out of one compartment into an adjacent one; and, in the other, dissolved substances pass from compartment to compartment. But which is which? Make notes on the term osmotic pressure. See *Rutishauser 1994, Figs 13.1 and 13.2.*

3. Another force is hydrostatic pressure. (We met both this and osmotic pressure in an earlier Unit of this Workbook. Can you remember where?) Make notes, together with diagrams, to illustrate the principle of hydrostatic pressure. (See *Rutishauser 1994, Fig. 13.3.*) Remember that water can only move out of one compartment into another if the walls separating them are permeable to water – in other words, the walls must be able to 'leak', and let water through. Not all membranes between fluid compartments in the body have the same permeability.

   As an example of lack of permeability think of some types of waterproof clothing. Mountaineers among you might remember the earliest clothing produced which, it was claimed, would keep rain out. It did; but you got just as wet inside because the material was impermeable to water vapour from perspiration as well as to rain. Newer clothing is designed to keep rain out, and to let through water vapour from the climber's body as well.

4. If you look back at the normal electrolyte levels you noted earlier, you'll see that the levels for sodium are far higher than those for potassium. Can you find out why this should be? (Look at *Rutishauser 1994, Table 13.1.*)

 COMMENTARY ON QUESTION 4

When serum sodium levels are measured in a laboratory, it's the extracellular levels that are measured rather than the intracellular. What does this suggest about the usual sites for sodium and potassium?

---

5. Find out about the mechanism by which sodium and potassium ions move from the inside of a cell to the outside, and vice versa. Sometimes this happens by diffusion – if there's a concentration gradient – but in other situations energy is used to bring about electrolyte movement. What name is commonly given to this mechanism? See *Rutishauser 1994, Fig. 13.6.*

SECTION 2: REGULATING ELECTROLYTE LEVELS
Time for completion: about 2 hours

This Section is complex, and your notes need to be clear and precise. One way of clarifying certain points is to draw a series of small diagrams illustrating, for example, the movement of one particular electrolyte at various areas of the nephron.

   Our starting point is the glomerular filtrate which contains different levels of the electrolytes sodium, potassium, chloride and calcium. (See *Rutishauser 1994, Table 8.1.*) My suggestion is that you concentrate initially on the movement of sodium and potassium ions; once they are clear in your mind, go back and look at the others.

6. How are sodium and potassium (then calcium and chloride) reabsorbed from the proximal convoluted tubule? Make notes for each of these electrolytes,

perhaps accompanied by a small diagram. See *Rutishauser 1994, Fig. 8.9*, for an illustration of the role of transporter proteins and diffusion.

7. Now make notes and diagrams to show the movement of the above electrolytes between the distal convoluted tubule and the capillaries. See *Rutishauser 1994, Fig. 8.10*. Read about the role of the hormone aldosterone in maintaining sodium and potassium levels. Aldosterone was mentioned in an earlier Unit of this Workbook. Remind yourself where it is produced and under what conditions.

8. When you've made notes about the movement of chloride and calcium ions from different parts of the tubule to the capillaries, find out what hormonal factors affect levels of calcium. See *Rutishauser 1994, p. 250*.

For all of Section 2 see the following references: *Hubbard & Mechan 1987, pp. 97–99 and 100–101; Rutishauser 1994, pp. 164 and 243; Marieb 1992, pp. 884–887; Guyton 1984, pp. 359–361; Hinchliff & Montague 1988, pp. 522–525*.

As usual these references are arranged in order of increasing complexity. My advice is not to make use of the Hinchliff & Montague until you've gained a good preliminary grounding. You might also then like to dip into *Guyton 1991, Ch. 27*. The opening paragraphs to this chapter (*p. 298*) are a useful introduction to how urine is concentrated in the tubules.

SECTION 3; REGULATING FLUID BALANCE
Time for completion: about 2 hours. Completing Sections 1 to 3 should take you about 1 'college day'.

9. Write notes on how water is reabsorbed from the filtrate in the proximal convoluted tubule. With the aid of diagrams, show how different parts of the loop of Henle and the distal convoluted tubule have differing permeability to water. How does the cellular construction achieve this?

 COMMENTARY ON QUESTION 9

Remember that the loop of Henle traverses kidney tissue from the cortex to the medulla. The interstitial fluid surrounding the tubules, the loop of Henle, and their capillaries has different concentrations of sodium chloride. Do you understand the terms:

- isotonic
- hypotonic
- hypertonic?

Water will move from one solution to a stronger solution, but only if the membrane separating them is permeable to water – hence the importance of grasping the differing permeabilities of the tubule and loop of Henle walls. If you find that your notes on this highly complex subject are becoming too wordy, do make use of a series of small, simple diagrams. Figure 4.5 is my attempt to show the movement of sodium chloride (NaCl) into and out of the various areas of the loop of Henle. I've omitted to label the parts of the loop that are impermeable to water – you can add that information to your own diagram(s). Similar diagrams could be constructed showing the movement of water or potassium.

My reason for suggesting these series of simple diagrams is that figures in many textbooks contain, for me at any rate, too much information; they are too detailed. So I try to reduce the amount of information to take in from each diagram, by producing a number of consecutive figures, rather like a series of frames making up a cine film.

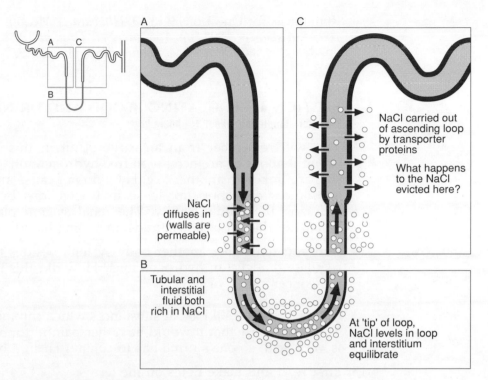

Fig. 4.5 Movement of NaCl in the loop of Henle.

10. What is the link between antidiuretic hormone (ADH) and the permeability of tubule walls?

11. Make notes on the vasa recta, its position and structure, and how it helps to maintain the osmotic differences between the cortex and the medulla of the kidneys. See *Rutishauser 1994, Fig. 8.4C*.

12. What is meant by the term autoregulation in relation to the kidney? What happens to urine production when blood pressure in the kidney arterioles either rises or falls? Make notes on how such changes are monitored by the body, and what responses are made by certain renal structures.

 FURTHER STUDY

It may be useful for you to read about how the body gains and loses water daily. *Table 13.2* in *Rutishauser 1994*, for example, shows how much fluid is lost per day in sweat and expired air. Under what circumstances might these losses increase?

On holiday in a very warm climate, if you maintained your usual fluid intake in the form of drinks, would your urine output rise or fall?

Sometimes people claim they have to visit the toilet more often in cold weather. One reason for their producing more urine could be an increase in their intake of hot drinks. Can you think of any other physiological reasons?

---

13. Where is the body's thirst centre situated, and what part does it play in the control of fluid levels?

References for Questions 9 to 13: *Hubbard & Mechan 1987, pp. 101–104 and 107–109; Rutishauser 1994, pp. 166 and 241; Guyton 1984, pp. 372–377; Marieb 1992, pp. 888–893; Hinchliff & Montague 1988, pp. 525–529.*

SECTION 4: REGULATING BLOOD AND URINE PH
Time for completion: about 1½ hours

You will remember from an earlier Unit in this Workbook that acidity is brought about by an increase in free hydrogen ions (H^+) in a fluid. Those ions that are linked to another (e.g. H_2O) do not cause an increase in acidity. Some substances can 'mop up' loose hydrogen ions by attaching to them, thus reducing the fluid's acidity. The renal system plays an important part in balancing the pH of body fluids, including blood and urine.

14. Can you remember another body system, besides the renal system, that can regulate the level of hydrogen ions? How did this body system act to reduce the concentration of H^+?

In this Section we'll look at substances which can mop up or buffer hydrogen. You will realise that it would be rather painful for a person to pass urine that is very acid, so one's urine has to contain H^+ in a buffered form.

15. First read and make notes on the terms:

    - acid
    - alkaline
    - base
    - buffer.

See *Rutishauser 1994, pp. 223–225.*

16. What is the meaning of the term pH? What exactly does pH measure, and what does the pH reading (e.g. pH = 6.5) signify? See *Rutishauser 1994, Table 12.1.*

17. What is the normal pH of blood, and what is its normal range (i.e. the range of pH that is compatible with life)? Compare the pH of blood to the pH of:

    - gastric acid
    - pancreatic juice.

See *Rutishauser 1994, p. 225.*

As you will have remembered earlier (Question 14) the respiratory system is able to make swift adjustments to the pH of the blood, by either hyperventilation or hypoventilation. You'll have recalled that hydrogen ions can be excreted 'disguised' as H_2O and CO_2 (water and carbon dioxide). The renal system is much slower at buffering hydrogen ions than the respiratory system, but it is an important route for H^+ to be either conserved or excreted, depending on the pH of the body.

18. It might be a good idea to revise the ways in which the respiratory system regulates pH – see *Rutishauser 1994, p. 228.* Then continue reading this chapter and make notes on:

    - how bases are excreted in the urine or conserved (e.g. hydrogen carbonate, hydrogen phosphate and sulphate). It's perhaps more useful to concentrate on HCO_3^- movement rather than the others, until you've attained a clear understanding of events in the tubules.

• how hydrogen ions are excreted or conserved. For example, what part do phosphates and ammonia play in the excretion of H^+?

COMMENTARY ON QUESTION 18

Draw simple diagrams for each part of this question, but don't try to cram too much information into one diagram. Show how each substance manages to pass from, for example, the tubular cells into the capillary. Is it by simple diffusion, or by the sodium–potassium pump, or by means of a transporter substance? Figure 4.6 is my attempt to construct a step by step diagram illustrating how H^+ is excreted in the urine in buffered form. You could use this as a 'template' for diagrams of your own, or adapt it. One small hint: when you draw a series of diagrams, keep the blood capillaries and the tubular fluids on the same respective sides of all your diagrams. This will help to clarify direction of movement of ions etc.

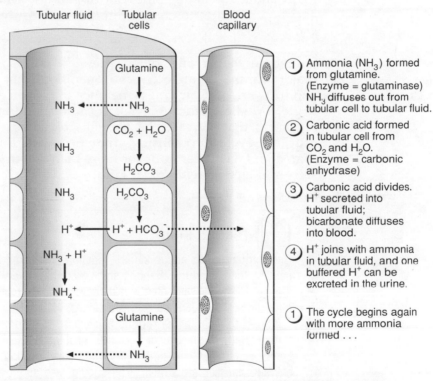

Fig. 4.6 Excretion of H^+ in the urine. (Adapted from Jennett 1989, Fig. 10.23.)

ACTIVITY

Find out how nurses test the urine of patients in hospital, in particular how the urine is tested for its pH. Special dipsticks can be used, in which colour changes denote the urine's pH. Another method is to use strips of blue and red litmus paper. If blue litmus paper turns pink when it is dipped in the specimen of urine, what does this denote?

When dipsticks are used, the instructions on the jar must be read and followed exactly. Usually, strict timings are given for each of the colour strips to be read. Failure to follow these times exactly will give inaccurate results.

19. How are bases and acids taken in by the body? Are they present, for example, in the diet; if so, in which foodstuffs? Is it possible for someone to include too much H^+ or HCO_3^- in their diet?

20. Find out the meaning of the terms acidosis and alkalosis. What levels of pH might be associated with each? How are they caused – find out about the difference between respiratory acidosis and metabolic acidosis. See: *Hubbard & Mechan 1987, p. 107; Rutishauser 1994, p. 234 and Fig. 12.10; Marieb 1992, pp. 921–923.*

SECTION 5: THREATS TO HOMEOSTASIS
Time for completion: about 1½ hours

In Question 12 you learned about responses made by the kidneys to a rise or fall in the blood pressure. Within certain limits, the kidneys adjust to changes in the renal blood flow which arise from changes in the systemic blood pressure. This is a good example of homeostasis.

21. Read about the juxtaglomerular apparatus: where it is found in the kidney, and how it reacts to changes in blood pressure. See *Rutishauser 1994, p. 245, and Fig. 13.8.*

Sometimes a person's blood pressure falls so low, for example following major trauma or an extensive operation, that his or her kidneys cease to produce urine. (Can you remember the medical term for non-production of urine?) Medical and nursing staff can respond in time to such a situation with measures that increase the patient's blood pressure – and with it the urine output. Severe hypotension combined with low urine production (e.g. less than about 35 ml per hour for more than 2 hours) should always be a warning sign of incipient acute renal failure.

 FURTHER STUDY

Once you're clear about kidney function, read about acute renal failure. Note that, usually, there is an initial anuric stage (with no or very little urine produced) followed by a polyuric stage (when plenty of dilute urine is excreted). You might try to work out the physiological significance of these stages.

Linked to renal failure (but perhaps more to chronic failure than acute) as a topic for further study is the use of a kidney machine – an artificial kidney. There are different types of machines, and different ways of dialysing (look up this term) the patient, but the overall principles are similar. You should read about the type of sterile fluid (dialysate) that is used in order to draw from the patient those electrolytes and waste products that he is unable to excrete for himself. If the patient is not excreting sufficient water, you could try to work out what type of dialysate could be used to draw off water from his body. See: *Rutishauser 1994, pp. 163 and 170; Marieb 1992, p. 892.*

Be warned that renal failure and renal dialysis are highly complex subjects. Be content, at this stage of your training, with a simple, clear overview of each. As you make your notes and construct your diagrams, be sure to make clear links with the physiology that you've already studied here.

You may remember the second patient profile I outlined briefly at the beginning of this Guided Study, that of a young child who'd had severe diarrhoea for 3 days. Go back and check the clinical features I gave about him.

22. Vomiting and diarrhoea are two common ways of losing great quantities of fluid from our bodies. In two columns, one for each condition, list the substances (as well as water) that will be lost from the body. Is prolonged vomiting likely to lead to a state of acidosis or alkalosis, and why? Which is diarrhoea likely to lead to, and again why? See: *Hubbard & Mechan 1987, p. 107; Guyton 1984, pp. 381–382; Rutishauser 1994, pp. 234 and 244.*

23. Bowel disturbances are very common in Third World infants, and many lives are lost through a condition as relatively simple to treat as diarrhoea. What do you think might be the best treatment for this condition, in order to correct the physiological inbalance?

24. The first patient profile I gave at the beginning of this Guided Study was of someone in fluid overload. This can occur in heart failure, in kidney failure (where fluid is not excreted) and if a patient's 'drip' is allowed to run into the vein too quickly. Drugs may be given to correct this fluid inbalance: they are called diuretics. (As well as giving a diuretic, what other treatment do you think might occur?)

Different types of diuretic act on different parts of the nephron. This means that a patient may not respond to the first drug prescribed by the doctor, but another may be very effective. Look up the following diuretics and make notes on how they work on the kidney:

- frusemide
- spironolactone.

Frusemide is called a loop diuretic – why is this? Note that this drug, used over a prolonged period, leads to excretion of potassium as well as water. It's usual therefore to prescribe a potassium supplement with frusemide. Spironolactone has a different physiological action and does not require potassium – it's referred to as potassium sparing.

For this question, use one of the pharmacology texts written for nurses, for example Hopkins 1992.

REFERENCE
Hopkins S J 1992 Drugs and pharmacology for nurses, 11th edn. Churchill Livingstone, Edinburgh

■ The urinary system (check quiz)

1. Write down a very brief overview of the function of the kidneys, as if explaining it to a care assistant on your ward.

2. Explain how filtration occurs in the glomerulus. Give three examples of substances that are filtered in health, and two examples, besides plasma proteins, of substances that are not filtered. Why is it important that plasma proteins remain in the blood, and don't (in health) enter the filtrate?

3. Explain how water and sodium are reabsorbed in the tubules and loop of Henle. How is sodium excreted in the urine, and how is this process controlled?

4. Mrs Vernon is involved in a serious car crash, and she loses a lot of blood.

    a. What effect will this blood loss have on her pulse and blood pressure? (This is revision of an earlier Unit.)
    b. What effect will this blood loss have on her production of urine?
    c. With your knowledge of physiology gained so far, what nursing observations do you think Mrs Vernon might need following her admission to hospital? Think about the changes in her cardiovascular and urinary systems, and decide which observations could be required to monitor the functioning of those important systems.

5. Give one example of an intracellular ion and one of an extracellular ion. Give the chemical symbol of each.

6. What does the term pH mean? What is the pH of blood?

7. What is meant by the terms acidosis and alkalosis? How might respiratory acidosis and metabolic acidosis be caused?

8. A patient suffers a cardiac arrest, and although resuscitation is begun immediately by nursing staff his heart doesn't start to beat for the first 10 minutes or so. Bearing in mind that he isn't breathing, do you think he will develop respiratory acidosis or alkalosis? Can you explain why?

9. Mrs Hadley, an elderly widow, has complained to her doctor of increasing breathlessness, and swelling of her feet, especially during the evening. After examining her, the doctor prescribes a diuretic.

    a. What does this term mean?
    b. Give two examples of diuretics and, choosing one of them, explain how it will help reduce Mrs Hadley's tissue oedema and improve her breathing.

10. Explain briefly the physiology of micturition. Distinguish between the roles of the autonomic and somatic nervous systems.

11. Besides micturition, how else is water excreted from the body?

Human continuity

Contents

UNIT 5

■ Reproduction (guided study)

Time for completion About 7 to 8 hours

Overall aim To explore the structure of the male and female reproductive systems, and relate these to the means by which an ovum can be fertilised by a sperm.

Introduction Today's youngsters allegedly know 'all about sex', and it's probably true that television, videos, and some daily newspapers have contributed to a greater awareness of certain sexual matters. However, the continued rates of unwanted teenage pregnancies, and the spread of sexually transmitted diseases, including HIV, might suggest that there are gaps in the sexual knowledge promoted by the media, as well as some inaccuracies in what appears. For example, there are grounds for suggesting that the tabloid representation of HIV and AIDS is not 100 per cent accurate.

Consequently, individuals continue to learn, and therefore promulgate, fiction rather than fact about sexual matters, such as the belief that a girl doesn't become pregnant the first time she has sex, or if she 'does it' standing up. Again, young men, brought up with media that promote women as ever available, continue to regard contraception as exclusively a woman's responsibility.

Without wishing to preach, may I suggest that nurses – leading lights in health education – are professionally bound to discover, and then promote, matters of sexual fact in order to counteract the sexual fantasies preferred, and widely spread, by some sections of the media.

This Workbook is concerned with physiological and, in part, psychological aspects of sexual activity. There is much to do with sex that is social and ethical, perhaps even spiritual, which is beyond its scope.

Background questions In this Guided Study, I begin by asking you to check your present understanding of certain terms and aspects of reproduction. Try to jot down your answers to the following questions without reference to other texts, so that you develop a good idea of your knowledge 'baseline'.

1. Write brief notes on your understanding of the menstrual cycle.

2. Conception is prevented in a number of ways. What are the contraceptive principles of:

   - the condom
   - the intra-uterine device (IUD)
   - the 'pill'.

3. What physiological reasons might there be for a couple failing to conceive? (Don't forget problems affecting the male.)

Specific questions **SECTION 1: MALE REPRODUCTIVE ORGANS**
Time for completion: about 1 hour

1. Draw a diagram showing the structure of the male genitalia and related organs. Start with the testes and scrotum, the urethra and penis, and the bladder. Then add the epididymis and the vas deferens. Then add the various glands of the male reproductive system:

   - prostate gland (be careful of the spelling)
   - seminal vesicles
   - bulbo-urethral glands.

   Note the position of the prostate gland in relation to the rectum. (If you can show the rectum on your diagram without making it seem cluttered, it would be useful to do so. A doctor can feel the prostate, and judge its size, by inserting a finger into the patient's rectum.)

Your diagram will provide you with an overview of the male genitalia. To help you construct your own diagram, refer to: *Rutishauser 1994, Fig. 34.1; Marieb 1992, Fig. 28.1.*

2. Make detailed notes (perhaps aided by a further diagram) on the structure of the penis. Note the tissues from which it is constructed, its blood supply, and the 'mechanics' by which it can become erect. Note the role of the autonomic nervous system here. See: *Rutishauser 1994, Fig. 34.5; Guyton 1984, Fig. 37.2; Marieb 1992, Fig. 28.4.*

   In which parts of the skin covering the penis are sensory nerve endings in greatest number?

3. Now we turn to the structure and function of the testes (sing. testis). Note the structure of the two testes within their loose sac of skin, and their rich blood supply via the spermatic cord. Draw a diagram showing a cross-section of a testis, with its seminiferous tubules and interstitial cells. Show also the coverings (tunica) of the testis, and how the inner covering forms many compartments or lobules, into which the seminiferous tubules are crammed. Show how the epididymis leaves the testis. See: *Rutishauser 1994, Fig. 34.2; Marieb 1992, Figs 28.2 and 28.3.*

4. Next we look at how sperm are produced. Make notes on where stem cells are found, and how they divide and mature into fully developed spermatozoa. (See *Rutishauser 1994, Fig. 34.3.*) What is the role of the Sertoli cells in this process? You should also make notes on the role in sperm production of these two hormones:

   - follicle stimulating hormone
   - testosterone.

Where are they produced? Besides affecting sperm production, testosterone has other functions. What are they?

 COMMENTARY ON QUESTION 4

You can address this question at differing levels. *Rutishauser 1994, Ch. 34*, gives a clear overview of the process of sperm production, though you'll have to dip into *Ch. 2* in order to understand the processes of cell division. *Marieb 1992, pp. 938–943*, goes into much more detail and you'd be as well to attain a sound understanding first before tackling this text. A good 'in between' text on sperm development is *Guyton 1984, pp. 611–614*, though the diagrams are, in this instance, a little less helpful than either Marieb's or Rutishauser's.

One of the functions of testosterone is the development and maintenance of the male secondary sex characteristics. What are these? In other words, what are the physical features, apart from the development of the external genitalia, that characterise maleness? See *Rutishauser 1994, Ch. 35*. On the role of testosterone, and other related hormones, you could also try: *Hubbard & Mechan 1987, pp. 332–334; Guyton 1984, p. 616*.

A related topic is that of changes in the male at puberty. You could discover what changes in, for example, hormone production occur at puberty, and how it is that the secondary sex characteristics mentioned above are brought about. Are there any psychological changes, as well as physical and physiological?

5. What part do the male genital glands (listed above in Question 1) play in the production of semen? What is the volume of semen produced at ejaculation? (Any answer can, of course, be only an average. Seminal fluid volume varies according to the frequency of ejaculation.)

6. What is the normal sperm count? Your answer will be in millions per millilitre. What happens to fully developed spermatozoa that are not ejaculated? (The physiology of ejaculation will be explored in Section 3 of this Guided Study.)

See *Jennett 1989, pp. 422–423* for an interesting discussion on spermatogenesis under the heading: 'Why so many?'.

 FURTHER STUDY

You might like to read about causes of male sexual problems, such as impotence, premature ejaculation, and sterility. You'll discover that only rarely are there causes that are purely physiological, and that psychological causes are common. See Bullock et al (1989) for a detailed discussion of male infertility and other sexual problems.

Enlargement of the prostate gland, which is fairly common in the elderly male, can cause obstruction to the flow of urine, leading to retention of urine, an extremely painful condition which must be treated swiftly. Read about changes in the prostate gland that can cause this enlargement. Note that only the minority of cases of hypertrophy are due to cancer.

Why is it, by the way, that the testes are situated somewhat vulnerably outside the main body of the male? In trying to answer this, you might also read about the development of the testes in the very young male child and into puberty.

SECTION 2: FEMALE REPRODUCTIVE ORGANS AND THE MENSTRUAL CYCLE
Time for completion: between 1½ and 2 hours

7. As in the preceding Section, begin by drawing a clearly labelled diagram of the female reproductive organs. Show:

- the uterus
- the ovaries
- the uterine tubes
- the vagina.

Label the different parts of the uterus, and show how these structures are maintained in position by ligaments.

You may like to draw a small-scale lateral view of the female organs (as in *Marieb 1992, Fig. 28.11*) to show their relationship to the bladder and rectum. A lateral view will also show the forward tilt of the uterus. What name is given to this position? See: *Wilson 1990, Fig. 15.3; Rutishauser 1994, Figs 34.7 and 34.8; Marieb 1992, Fig. 28.11.*

8. Describe the position of the uterine tubes within the pelvic cavity. What is their position in relation to the peritoneum and the ovaries? What other name is sometimes given to the uterine tubes?

What type of tissue both forms and lines these uterine tubes and, acting together, what is their specialised function?

9. Describe the position of the ovaries, and explain (perhaps with the aid of a diagram) how they are maintained in position within the pelvic cavity.

See *Wilson 1990, Fig. 15.4*, and the preceding references from Rutishauser 1994 and Marieb 1992 for Questions 8 and 9.

10. You've already included the uterus in your earlier diagram of the female reproductive organs. Check that you're sure of the position of the uterus, and the name given to this position. Now describe the different parts of the uterus, including its body, and the cervix. Make notes on the type of tissues lining the uterus and cervix. Do you understand what the terms myometrium and endometrium mean?

11. Similarly, describe the tissues that form the vagina. What is the function of the cells lining the vagina? What sort of muscle cells are found in the vagina? How is the vagina lubricated? Your answer should include reference to the pH of the vagina, and how that pH is formed.

    What structures form the pelvic floor, and what is their connection with the vagina? (See *Rutishauser 1994, Fig. 34.9*.)

12. Draw a diagram showing the female external genitalia. (See *Rutishauser 1994, Fig. 34.10*.) Make notes explaining the secretions produced, and the areas that are most liberally supplied with sensory nerve endings.

 FURTHER STUDY

Your diagram should show how closely situated are the vagina, the urethra, and the anus. You may already know that urinary tract infections are commoner in the female than in the male – you may like to find out why, bearing in mind the proximity of the three structures already mentioned.

As with the male reproductive system discussed earlier, you'll find that Marieb 1992 provides greater detail than Rutishauser 1994. It's a good idea, then, to read through *Rutishauser Ch. 34, pp. 558–560,* first in order to obtain a clear overview of the structure of the female reproductive system before turning, if you wish, to *Marieb 1992, pp. 946–952*. Note that we'll study the structure of the breast in Section 5 of this Guided Study.

We conclude this Section by looking at the complex series of events that make up the menstrual cycle. First, check the notes you made to review your knowledge at the beginning of this Guided Study. Were there any areas that were a little problematic for you?

13. Write notes on the development of ova and the process of ovulation. You should include a description of the hormones influencing this process, where they are produced, and the changes that occur in the female reproductive organs, for example in the endometrium and vagina.

    Make a small diagram demonstrating the main features of ovulation. This will eventually contribute to a larger diagram showing all stages of the menstrual cycle, which you'll build up in stages as another 'series diagram'.

    Once again, read through *Rutishauser 1994, p. 566,* first, before tackling the more complex sections in: *Marieb 1992, pp. 954–959; Guyton 1984, pp. 617–620*.

    As you write your notes, you should attempt to highlight differences between the development of sperm in the male, and ova in the female.

14. The menstrual cycle occurs in a number of stages, but it is unfortunate for the nursing student that different texts give differing names to these stages. I suggest you read through *Rutishauser 1994, Ch. 34, pp. 567–568*, where the menstrual cycle is discussed in the following three stages:

    - menstrual stage
    - follicular stage
    - luteal stage.

    Under each heading, changes are described in various organs of the female reproductive system. Don't just describe these changes, but also discover their purpose. You should note how, and from where, the major hormones oestrogen and progesterone are produced, and any feedback mechanism controlling their production.

    In which stage is the yellow body produced, and what is its purpose? (There's another name for this structure that you should note.)

    Construct one or more small diagrams to illustrate the main changes (e.g. in the structure of the endometrium) throughout these three stages. These diagrams will contribute to your large final diagram.

15. Now describe the associated changes that occur throughout the menstrual cycle. *Rutishauser 1994, p. 568*, lists these under three headings:

    - breast tissue
    - body temperature
    - mood.

    What is the significance of body temperature for:

    - couples having difficulty in conceiving
    - couples unwilling to use artificial methods of contraception?

    On your diagram(s) try to show which parts of the menstrual cycle are best for intercourse:

    - if conception is the aim
    - if prevention of conception is the aim.

    See *Guyton 1984, pp. 624–625*.

 COMMENTARY ON QUESTION 15

This is more of a personal comment than a helpful hint on how to answer the question, and it is provoked by the inclusion of 'mood' in the list of changes associated with the menstrual cycle. It is perhaps male ignorance that continues to regard mood changes as a female quirk, something they 'put on'. In my own nurse training, male students were not permitted to attend lectures on female reproductive anatomy and physiology; consequently some of us continued with our lay knowledge that amounted to little more than assumption or fiction. No wonder some of us regarded female sexuality as a mystery and, especially, contraception as none of our business. There is no doubt that psychological problems associated with both the menstrual cycle and pregnancy itself can be severe. Similarly, psychological preparation for women undergoing certain types of surgery is at least as important as physical preparation. The inclusion of 'mood' in the list of problems is not done just for the sake of completeness. It is a very real factor in many women's lives. All of us – male as well as female nurses – should be aware of this.

Before finishing this Section, compare the large-scale diagram you've been building up in stages with one of the following: *Wilson 1990, Fig. 15.10; Rutishauser 1994, Fig. 34.14; Marieb 1992, Fig. 28.19.*

 Have a short break here. A more substantial break is planned for the end of Section 3.

SECTION 3: SEXUAL ACTIVITY
Time for completion: about 1 to 1½ hours

Physiologically, the point of sexual activity is to bring sperm and mature ovum together. Obviously, describing sex in such stark terms omits much that is important in human relationships, so it is as well to be clear from the outset of this Section that we'll be restricting our coverage of sexual activity to sexual intercourse, and the physiological changes that occur in both male and female partners.

16. During sexual arousal, vasocongestion occurs in both males and females, leading to erection of the penis and hardening of the clitoris (and other tissues). Find out how these changes occur. Make notes on the types of psychological and physical stimulation that can lead to vasocongestion. What physiological events lead to erection of the penis, and how is the erection maintained?

 COMMENTARY ON QUESTION 16

Stimulation need not be of a strictly sexual nature, such as touching erogenous zones. Some people can be aroused by other stimulants such as sights, sounds, and smells.

Vasocongestion can occur as a spinal reflex, but there is almost always nervous input from the higher centres. Show these inputs from different levels in a simple diagram.

---

17. Describe the fluid secretions that occur during sexual arousal in both men and women. What are their main purposes?

18. Make detailed notes on the events leading up to, and at, orgasm in both male and female. (You'll realise that there are both similarities and differences between the sexes here.) Your notes should include:

   ● a brief description of the psychological experiences before, during, and immediately after orgasm
      ● emission and ejaculation in the male
      ● smooth muscle activity in the female related to the female orgasm
      ● skeletal muscle activity in both male and female
   ● changes in heart rate, blood pressure, and respirations leading up to, during, and immediately after orgasm, in both sexes.

19. What is the latent period in the male following ejaculation? How, in this respect, does the male orgasm differ from the female?

20. Now make notes on the journey of sperm, following ejaculation, into the uterine tubes. Look back at your earlier notes and recall how many sperm, approximately, are deposited into the vagina at ejaculation. What is the function of some of the substances in seminal fluid in helping the sperm reach their destination? (Fertilisation is studied at the beginning of Section 4.)

Your principal reference for Section 3 should be *Rutishauser 1994, pp. 562–566*, because this text (sensibly in my view) covers both male and female responses and activities in one section. (Another is *Hinchliff & Montague 1988, pp. 627–629*.) Most other textbooks I've seen provide information in separate sections for male and female.

You could also look at: *Jennett 1989, pp. 425–426 and 430–431; Marieb 1992, pp. 943–945 and 961; Guyton 1984, pp. 614–615 and 620–621; Guyton 1991, pp. 890–891 and 911–912.*

 FURTHER STUDY

Look back at your response to Background Question 2, and in particular your description of the contraceptive principle of the condom. You'll realise now (and perhaps did at the beginning of this Guided Study) that the condom works by providing a physical barrier that, if all goes well, sperm cannot cross.

Are there any other forms of contraception that similarly provide a barrier between sperm and ovum? What are the advantages and disadvantages of each form of contraception? (For a discussion of contraception see *Marieb 1992, pp. 962–963*.)

One form of contraception – not one to be recommended – is coitus interruptus, where the penis is withdrawn from the vagina immediately before ejaculation. The psychological problems of this method are doubtless obvious, but what are the physiological drawbacks? In other words, why does coitus interruptus often not work in preventing conception? (Think back to your studies of the stages of sexual arousal and male orgasm.)

 Have a long break here. So far this Guided Study may well have taken the whole of a morning, or even a whole day.

SECTION 4: PREGNANCY
Time for completion: about 2 to 3 hours

21. We left Section 3 with the ejaculated sperm journeying through the cervix and into the uterus. Whereabouts in the female reproductive system does fertilisation usually occur?

22. What are the changes in sperm and ovum that help make fertilisation more likely? See: *Rutishauser 1994, Fig. 34.15 and p. 568; Guyton 1984, pp. 628–629; Marieb 1992, pp. 975–979.*

23. How is the full complement of chromosomes in the fertilised ovum achieved? How is the sex of the embryo determined? (There is a Commentary on this question on p. 180.)

COMMENTARY ON QUESTION 23

You'll need to recall how many pairs of chromosomes are carried by both the sperm and ovum, and how these merge into the final complement in the fertilised egg, thus providing all the genetic determinants of the newly created individual. See: *Rutishauser 1994, Fig. 35.2, p. 584; Mackenna & Callander 1990, pp. 20–21.*

24. Describe what happens to the fertilised egg (the zygote) for the first few days of its development until implantation into the uterine wall. How can the cells of the (eventual) placenta be distinguished from the cells of the (eventual) embryo? See: *Rutishauser 1994, Fig. 34.16; Mackenna & Callander 1990, pp. 21 and 204–206.*

    Please note that the diagrams shown in Mackenna & Callander 1990 are summaries of events. You should read, and ensure you understand, the descriptions given in Rutishauser or your other preferred text before checking your knowledge with these summaries. For a more complex description than Rutishauser 1994 see: *Guyton 1984, pp. 628–630; Marieb 1992, pp. 979–982* (including some useful electron micrographs).

25. Make notes on how the placenta develops, from certain of the early few cells to a large organ. What is its function? Your notes should include a description of the provision of nutrients from the mother's circulation via the placenta, and the disposal of waste products. (Leave the hormonal production of the placenta to the next question.) See: *Rutishauser 1994, Fig. 34.17 and p. 570; Guyton 1984, pp. 630–633; Marieb 1992, pp. 982–984.* Then check your notes (and diagram(s)) against *Mackenna & Callander 1990, p. 207.*

26. You'll have discovered that the placenta acts as a kind of intensive care unit for the growing fetus, but it has another function: that of producing certain hormones. Read about this, and make notes on each of the hormones produced. You may find that constructing your own line diagrams will help your understanding. See *Rutishauser 1994, p. 572.*

 FURTHER STUDY

You've already made notes on how certain substances (such as oxygen and other nutrients) can pass from the maternal circulation to the fetus. What about some unwanted substances? Find out what advice is given to expectant mothers regarding, for example:

- cigarette smoking
- alcohol
- medicines.

It's easy to guess that mothers will be advised to stop smoking – but why? What effect might smoking have on the growing fetus?

27. As the fetus develops so does its circulation and nervous system. To read about these, you'll need to turn to *Ch. 35* of *Rutishauser 1994*. See also: *Guyton 1984, pp. 636–637; Jennett 1989, pp. 409–413*.

    Although *Marieb 1992, pp. 984–992*, provides a good deal of information, as well as some excellent diagrams, it is as well to gain a sound understanding before tackling this text.

28. The developing fetus is bound to have considerable effects on the mother carrying it, so that she must make many physical and physiological adjustments. You may already know that iron supplements are advised in pregnancy, and you'll probably know enough to work out why.

    Make notes on the adjustments made by the pregnant mother. *Rutishauser 1994* organises these under a series of headings that you may find useful – see *pp. 572–574*. You could also look at: *Guyton 1984, pp. 637–638; Marieb 1992, pp. 992–994*.

    Here are two questions just to help you check that you've studied this topic sufficiently.

    - Why does the mother sometimes suffer from constipation during pregnancy?
    - Why is there fluid retention during pregnancy?

    You might also like to discuss within your student group the problem of food fads or cravings that sometimes occur during pregnancy. What examples does the group know about? Why do you think these cravings occur? Also, what adjustments to the mother's diet should be made during pregnancy? (See *Table 34.2* in *Rutishauser 1994, p. 575*, which shows the recommended nutritional intake in pregnancy.)

SECTION 5: DELIVERY AND LACTATION
Time for completion: about 1 hour

The aim of this final Section is to develop a basic understanding of the physiological events leading up to, and during, birth, and of breast feeding. Please bear in mind that the level at which we study these subjects will be less than that required by student midwives.

29. What is the average length of time, in weeks, for pregnancy to last? What is the starting point for this length of time?

30. One quick revision point: of what type of muscle is the uterus composed? (You'll immediately rule out cardiac muscle, but what of the other two types – striped and smooth?)

    Describe the weak uterine muscle contractions that occur early in pregnancy. (Don't confuse these with the contractions that are a sign of imminent birth.) What causes these weak contractions?

    What changes in this uterine muscular activity occur as pregnancy proceeds? Again, what causes these contractile changes? (You'll need to discover the changes that occur in the ratio of certain hormones. See *Fig. 34.22* in *Rutishauser 1994*, for a summary of hormonal changes.)

31. What skeletal adjustments are made within the female body as pregnancy nears its final stages? (See *Rutishauser 1994, p. 576*.)

FURTHER STUDY

Once you've discovered the answer to Question 31, find a skeleton in your college classrooms and have a look at the bones of the pelvis. Try to relate what you see to the answer you've written down for Question 31. Incidentally, can you distinguish between a male and female skeleton?

32. In preparation for the birth of the baby, what changes happen to the cervix, and how are these changes brought about?

33. Draw a diagram showing the normal position of the baby within the uterus just before birth. You may also like to make very brief notes on other positions (some of which cause great problems for mother, baby, and midwife). See: *Guyton 1984, Fig. 38.10; Marieb 1992, Figs 29.16 and 29.17.*

34. What happens to the placenta following the baby's birth? How does the body prevent potentially severe blood loss at birth?

FURTHER STUDY

As most mothers will confirm, giving birth can be very painful. Read about the different methods of pain relief available during delivery, and apply your findings to the physiology you've learned here. For example, an intramuscular injection of pethidine will reduce the pain experienced by the mother, but are there any disadvantages of using this drug?

For a more detailed description of parturition than either Rutishauser 1994 or Guyton 1984, see *Marieb 1992, pp. 994–996.*

35. What is the meaning of the term puerperium? What physiological adjustments occur in the mother's body following the birth of her child? (Some of these changes depend on whether breast feeding occurs; other changes happen independently of this.)

36. Now we consider breast feeding and milk production. First draw a diagram of the breast showing the presence of glandular tissue and milk ducts. See: *Rutishauser 1994, Fig. 34.11; Marieb 1992, Fig. 28.15.*

   Make brief notes on the hormonal influences that bring about changes in the female breast at puberty. (Use earlier references in this Guided Study for this question, as well as *Ch. 35* in *Rutishauser 1994;* but see also the summary provided in *Mackenna & Callander 1990, p. 213.*)

37. Review your earlier findings of changes to the breasts during the menstrual cycle.

38. Milk production in the breasts occurs in greatly increased quantities after parturition. Explain the hormonal balance which allows this to happen.

39. Similarly explain what hormones are stimulated when the baby suckles at the mother's breast. (See *Rutishauser 1994, Fig. 34.24 and p. 580.*) You may find a

simple flow diagram will help you keep track of the action of hormones during and after parturition.

40. Compare the composition of maternal milk, cow's milk and manufactured baby milk.

    For the first two, see *Rutishauser 1994, Table 34.3*. You'll need to do a little investigation, either in the literature or in shops selling manufactured milk for the third.

41. 'Breast is best' is a slogan often found in leaflets for nursing mothers. Find out why this is so. Your answer should cover the case for breast feeding in terms of:

    - psychological benefits
    - nutritional benefits, and
    - benefits in protection against infection.

    In order to achieve a balanced answer, however, try to read the literature provided for mothers by the manufacturers of baby milk. You'll find the advantages of using manufactured milk explained there, and simple instructions on how feeding should be carried out safely. You'll probably find that the manufacturers often, very honestly, concede that 'breast is best' in most cases.

REFERENCE
Bullock N, Sibley G, Whitaker R 1989 Essential urology. Churchill Livingstone, Edinburgh

■ Ageing (suggested reading)

It is all too easy to generalise about the process of growing old. Sometimes we think that deterioration of memory, posture, muscular strength, skin condition, and control of excretion are inevitable consequences of old age. Such changes can and do occur, but we probably all know of some retired people whose energy and liveliness put us to shame.

Not all physiology texts contain separate chapters or sections dealing with the ageing process. In Guyton 1991, for example, you'll have to make good use of the index to find references to ageing and various body systems.

So my principal recommendation here is *Rutishauser 1994, Ch. 36*, with, additionally, *Jennett 1989, Ch. 17*.

Read about how cell death occurs; and progress to how ageing can affect the various systems of the body. Why, for example, can an old person's skin become wrinkled? Why do some old people complain that they have lost height. What type of memory loss can occur because of ageing (leaving aside problems related to pathology)?

Both *Jennett 1989* and *Rutishauser 1994, Ch. 37*, usefully discuss the meaning of death. Death used to be declared when a person stopped breathing (hence the practice of holding a mirror in front of the victim's mouth, to see if it misted). Now we know that death can be more complex. What precisely is meant by brain death?

How has the need for organs for transplantation complicated the diagnosis of death? If this subject interests you, you could find out your own hospital's policy regarding the provision of organs for transplantation, and the means by which the donor and his or her relatives are protected against abuse.

■ Reproduction and ageing (check quiz)

1. Give an overview of the functions of oestrogen and progesterone, showing how the balance of these hormones changes through the menstrual cycle.

2. What role is played by testosterone during and after puberty in the male?

3. Sandra and Brian have been trying for a family for nearly 2 years, but without success. What possible physiological reason might there be for this failure to conceive?

   At what stage of the menstrual cycle might it be best for intercourse to occur if conception is desired? How could taking Sandra's temperature daily help to fix this point in the cycle?

4. As a school nurse visiting a class of boys (average age about 15 to 16 years) you have been asked to explain to them the benefits of using condoms during intercourse. What main points would you want to cover in your talk? (You may wish to extend your talk beyond physiological matters.)

5. Mr Edwards, aged 78 years, has an average blood pressure of 170/100 mmHg (as recorded every week by his district nurse). Is this reading normal or high for a man of his age? What physiological changes could have led to a blood pressure of this level?

6. Why might an elderly person be particularly vulnerable to hypothermia during winter? Begin your answer with the main physiological changes of ageing, but feel free to extend your answer to other areas. What advice might a district nurse give an old person living alone, to prevent hypothermia?

Index